Prepared in cooperation with the U.S. Department of Energy Office of Environmental Management, National Nuclear Security Administration, Nevada Site Office, under Interagency Agreement DE-NA0001654

Conceptualization of the Predevelopment Groundwater Flow System and Transient Water-Level Responses in Yucca Flat, Nevada National Security Site, Nevada

Scientific Investigations Report 2012–5196

U.S. Department of the Interior
U.S. Geological Survey

Conceptualization of the Predevelopment Groundwater Flow System and Transient Water-Level Responses in Yucca Flat, Nevada National Security Site, Nevada

By Joseph M. Fenelon, Donald S. Sweetkind, Peggy E. Elliott, and Randell J. Laczniak

Prepared in cooperation with the U.S. Department of Energy Office of Environmental Management, National Nuclear Security Administration, Nevada Site Office, under Interagency Agreement DE-NA0001654

Scientific Investigations Report 2012–5196

U.S. Department of the Interior
U.S. Geological Survey

U.S. Department of the Interior
KEN SALAZAR, Secretary

U.S. Geological Survey
Marcia K. McNutt, Director

U.S. Geological Survey, Reston, Virginia: 2012

For more information on the USGS—the Federal source for science about the Earth, its natural and living resources, natural hazards, and the environment, visit *http://www.usgs.gov* or call 1–888–ASK–USGS.

For an overview of USGS information products, including maps, imagery, and publications, visit *http://www.usgs.gov/pubprod*.

To order this and other USGS information products, visit *http://store.usgs.gov*.

Suggested citation:
Fenelon, J.M., Sweetkind, D.S., Elliott, P.E., and Laczniak, R.J., 2012, Conceptualization of the predevelopment groundwater flow system and transient water-level responses in Yucca Flat, Nevada National Security Site, Nevada: U.S. Geological Survey Scientific Investigations Report 2012-5196, 61 p.

Contents

Figures

Plates

Plates 1–4 are available at *http://pubs.usgs.gov/sir/2012/5196/*.

Conversion Factors and Datums

Conversion Factors

Inch/Pound to SI

Multiply	By	To obtain
Length		
inch (in)	2.54	centimeter (cm)
foot (ft)	0.3048	meter (m)
mile (mi)	1.609	kilometer (km)
Volume		
acre-foot (acre-ft)	1,233	cubic meter (m^3)
Flow rate		
acre-foot per year (acre-ft/yr)	1,233	cubic meter per year (m^3/yr)
gallon per minute (gal/min)	0.06309	liter per second (L/s)
inch per year (in/yr)	25.4	millimeter per year (mm/yr)
Hydraulic conductivity		
foot per day (ft/d)	0.3048	meter per day (m/d)
Hydraulic gradient		
foot per mile (ft/mi)	0.1894	meter per kilometer (m/km)

Temperature in degrees Fahrenheit (°F) may be converted to degrees Celsius (°C) as follows:

°C=(°F-32)/1.8

Datums

Vertical coordinate information is referenced to the National Geodetic Vertical Datum of 1929 (NGVD 29).

Horizontal coordinate information is referenced to the North American Datum of 1983 (NAD 83).

Altitude, as used in this report, refers to distance above the vertical datum.

Conceptualization of the Predevelopment Groundwater Flow System and Transient Water-Level Responses in Yucca Flat, Nevada National Security Site, Nevada

By Joseph M. Fenelon, Donald S. Sweetkind, Peggy E. Elliott, and Randell J. Laczniak

Abstract

Contaminants introduced into the subsurface of Yucca Flat, Nevada National Security Site, by underground nuclear testing are of concern to the U.S. Department of Energy and regulators responsible for protecting human health and safety. The potential for contaminant movement away from the underground test areas and into the accessible environment is greatest by groundwater transport. The primary hydrologic control on this transport is evaluated and examined through a set of contour maps developed to represent the hydraulic-head distribution within the two major aquifer systems underlying the area. Aquifers and confining units within these systems were identified and their extents delineated by merging and analyzing hydrostratigraphic framework models developed by other investigators from existing geologic information. Maps of the hydraulic-head distributions in the major aquifer systems were developed from a detailed evaluation and assessment of available water-level measurements. The maps, in conjunction with regional and detailed hydrogeologic cross sections, were used to conceptualize flow within and between aquifer systems.

Aquifers and confining units are mapped and discussed in general terms as being one of two aquifer systems: alluvial–volcanic or carbonate. The carbonate aquifers are subdivided and mapped as independent regional and local aquifers, based on the continuity of their component rock. Groundwater flow directions, approximated from potentiometric contours, are indicated on the maps and sections and discussed for the alluvial–volcanic and regional carbonate aquifers. Flow in the alluvial–volcanic aquifer generally is constrained by the bounding volcanic confining unit, whereas flow in the regional carbonate aquifer is constrained by the siliceous confining unit. Hydraulic heads in the alluvial–volcanic aquifer typically range from 2,400 to 2,530 feet and commonly are elevated about 20–100 feet above heads in the underlying regional carbonate aquifer. Flow directions in the alluvial–volcanic aquifer are variable and are controlled by localized areas where small amounts of water can drain into the regional carbonate aquifer. These areas commonly are controlled by geologic structures, such as Yucca fault. Flow in the regional carbonate aquifer generally drains to the center of the basin; from there flow is to the south-southeast out of the study area toward downgradient discharge areas. Southward flow in the regional carbonate aquifer occurs in a prominent potentiometric trough that results from a faulted zone of enhanced permeability centered about Yucca fault. Vertical hydraulic gradients between the aquifer systems are downward throughout the study area; however, flow from the alluvial–volcanic aquifer into the underlying carbonate aquifer is believed to be minor because of the intervening confining unit.

Transient water levels were identified and analyzed to understand hydraulic responses to stresses in Yucca Flat. Transient responses have only a minimal influence on the general predevelopment flow directions in the aquifers. The two primary anthropogenic stresses on the groundwater system since about 1950 are nuclear testing and pumping. Most of the potentiometric response in the aquifers to pumping or past nuclear testing is interim and localized. Persistent, long-lasting changes in hydraulic head caused by nuclear testing occur only in confining units where groundwater fluxes are negligible. A third stress on the groundwater system is natural recharge, which can cause minor, short- and long-term changes in water levels. Long-term hydrographs affected by natural recharge, grouped by similar trend, cluster in distinct areas of Yucca Flat and are controlled primarily by spatial differences in local recharge patterns.

Introduction

Yucca Flat is in the northeastern part of the Nevada National Security Site (NNSS) in southern Nevada (fig. 1). During the period 1951–1992, hundreds of underground nuclear tests were conducted in the Yucca Flat area (U.S. Department of Energy, 2000b); many of these tests likely introduced radionuclide contaminants into the groundwater (Laczniak and others, 1996, table 4).

The potential for subsurface transport of radionuclides in Yucca Flat is of concern and interest to the U.S. Department of Energy and to other regulatory Federal and State agencies (State of Nevada and others, 1996). Numerical models are being developed to simulate the flow of groundwater and

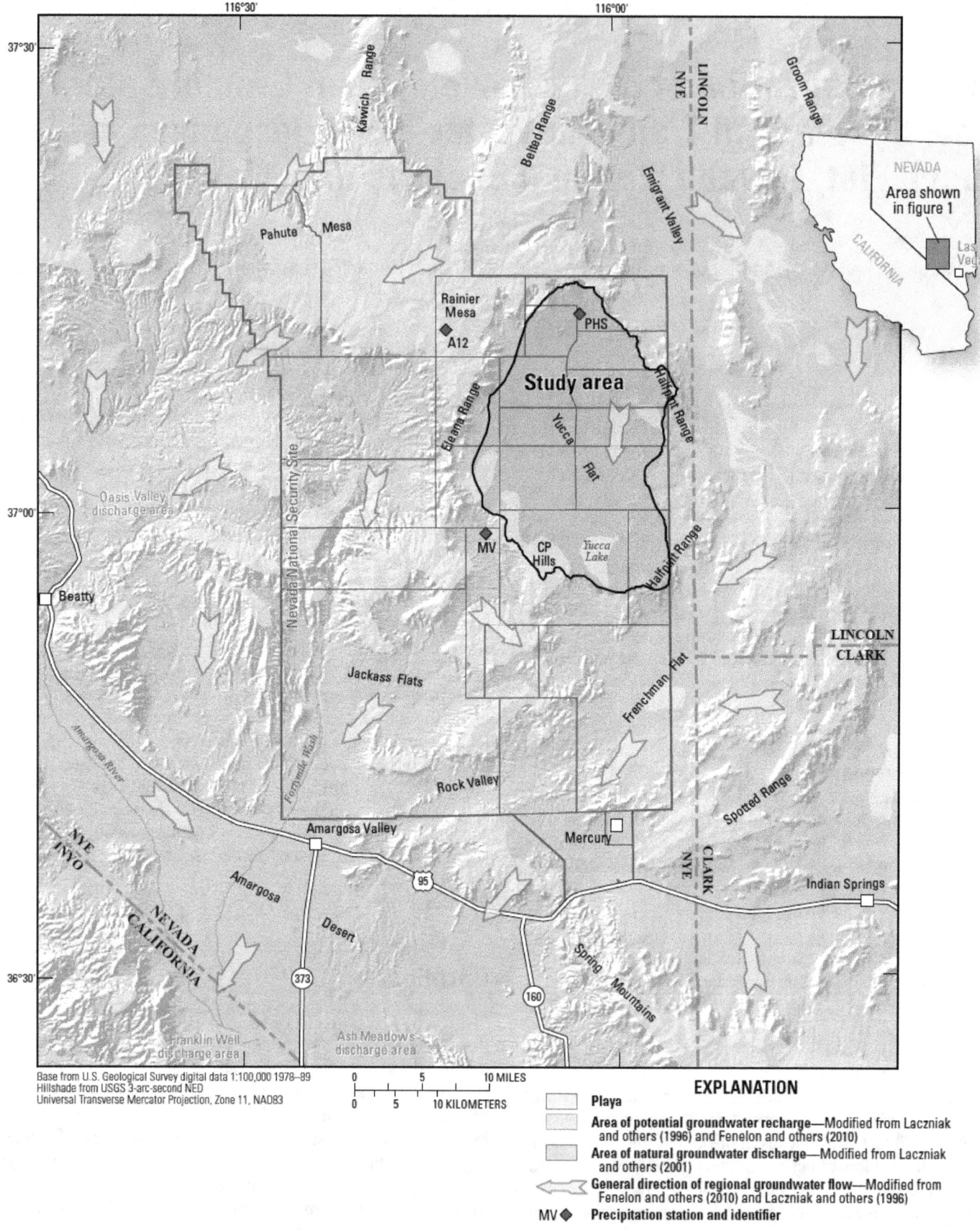

Figure 1.   Geographic and hydrologic features in the area near Yucca Flat, Nevada National Security Site.

the transport of contaminants away from areas in Yucca Flat (U.S. Department of Energy, 2000a). As part of this effort, wells have been drilled over the past 20 years and a number of geologic studies have been completed to better characterize the geology, hydrology, and geochemistry of the subsurface environment (Phelps and others, 1999; Fenelon, 2005; Halford and others, 2005; Bechtel Nevada, 2006; Farnham and others, 2006; Stoller-Navarro Joint Venture, 2006; Stoller-Navarro Joint Venture and others, 2007; Sweetkind and Drake, 2007a, b; Carle and others, 2008; McNab, 2008; Pawloski and others, 2008; Tompson, 2008; Asch and others, 2009; Stoller-Navarro Joint Venture, 2009; Drellack and others, 2010). Geologic data obtained from these and other sources, and insights gained from these investigations have been integrated spatially to create three-dimensional hydrostratigraphic framework models (HFMs) of the regional and local hydrostratigraphy (Faunt and others, 2004; Bechtel Nevada, 2006; National Security Technologies, 2007a). These HFMs portray the groundwater flow system as a complex series of aquifers separated by confining units. Flow and transport models founded on these geologic frameworks are being used to simulate near- and far-field transport of contaminants introduced into the groundwater flow system by underground testing and to formulate decisions regarding the management of these contaminants.

The direction and rate of subsurface transport away from former underground testing areas is controlled, in part, by the hydraulic-head gradient. The difference in hydraulic head over a given area defines the gradient and describes the groundwater flow potential. The spatial distribution of hydraulic heads throughout the NNSS region typically has been conveyed on maps by a single set of generalized potentiometric contours. Most of these maps are regional in scope (Fenske and Carnahan, 1975; Waddell and others, 1984; Laczniak and others, 1996; D'Agnese and others, 1998; Harrill and Bedinger, 2004), but some are focused on Yucca Flat (Doty and Thordarson, 1983; Hale and others, 1995). Potentiometric-surface maps based on a multiple-aquifer conceptual model are few and include regional-scale maps of the NNSS area that show contours of hydraulic heads in rocks of Cenozoic and pre-Tertiary age (Winograd and Thordarson, 1975; Fenelon and others, 2010), and maps of the Rainier Mesa and Shoshone Mountain area that show contours of heads in a volcanic aquifer and in local and regional carbonate aquifers (Fenelon and others, 2008).

Maps that portray the hydraulic-head distribution in geologically complex areas within the NNSS as a single set of contours discount vertical flow components and generalize the subsurface geology as one continuous, regionally extensive two-dimensional flow system. In actuality, as is indicated by published hydrostratigraphic framework models, the groundwater flow system is made up of multiple aquifers that are hydraulically separated by confining units. The degree of hydraulic separation depends on the hydraulic properties of the intervening confining rock. Hydraulic separation of the aquifers in Yucca Flat creates multiple, semi-independent flow systems, in which the direction and rate of groundwater flow

is unique and is controlled largely by the head gradients within each aquifer. The presence of separated shallow and deep groundwater flow systems in Yucca Flat has been known for a long time (Winograd and Thordarson, 1975) but was never tied into map-view concepts or integrated into a detailed, three-dimensional analysis. Furthermore, hydraulic gradients have been altered by transient effects such as pumping and underground nuclear tests (detonations). Successful simulation and accurate forecasts related to radionuclide transport require an understanding of the flow direction and rate within individual aquifers, particularly in the aquifer or aquifers that are susceptible to radionuclide contamination. This report provides a compilation and analysis of water-level data; presents potentiometric contour maps that define hydraulic gradients in the major aquifers in Yucca Flat; and describes a conceptual model of groundwater flow in the major aquifer systems. This is accomplished through a temporal and spatial analysis of water levels, and a thorough integration of existing geologic, hydrologic, and geochemical information.

Purpose and Scope

The purpose of this report is to delineate the major aquifers and confining units beneath Yucca Flat, to define and describe the likely direction of groundwater flow in each of the aquifers under predevelopment conditions, and to document and describe transient effects on water levels and groundwater flow caused by groundwater withdrawals, underground nuclear tests, and variations in recharge.

Predevelopment groundwater flow directions are determined by constructing potentiometric surface maps. The maps are designed to delineate the spatial extent of the major aquifer systems and describe flow within and between the aquifers in the multi-aquifer flow system underlying Yucca Flat. Predevelopment conditions assume equilibrium or a near equilibrium state in the groundwater flow system prior to any major changes prompted by human intervention, such as pumping and nuclear testing. Maps and their component hydraulic heads can be used as calibration targets for flow models and can help identify likely groundwater flow paths.

The purpose of the transient-effects analysis is twofold. First, the analysis is used to identify which water-level measurements collected during periods of onsite groundwater activity (pumping and testing) actually represent predevelopment conditions. This is especially important in confining units where large hydraulic gradients can result from nuclear testing. Second, the observed hydraulic response to transient effects helps to conceptualize the predevelopment flow system and also provides an understanding of the magnitude of these effects on this system and ultimately on the transport of contaminants.

The report summarizes well-construction and water-level data acquired from 229 wells used to develop potentiometric contours and analyze transient flow. The aquifers and confining units associated with the open interval or intervals of a specific well are based on hydrostratigraphic interpretations published in three-dimensional framework models that detail the hydrogeology of the study area (Bechtel Nevada, 2006; National Security Technologies, 2007a). Hydrostratigraphic interpretations and well-construction and water-level data can be displayed using interactive spreadsheets included as appendixes in the report. All water levels are flagged to indicate their likelihood of representing predevelopment conditions or having been affected by nuclear testing or pumping.

This report consists of four major sections to describe the groundwater flow in the aquifer systems in Yucca Flat. The section "Hydrogeologic Framework" describes the extent and hydrologic characteristics of the important geologic structures and the primary aquifer and confining unit types. That section is followed by a discussion of predevelopment flow in the aquifer systems and the interaction of flow between systems. The third section discusses transient stresses and their effects on flow by examining hydrographs and describing the overall effect of human stresses on the predevelopment flow system. The fourth section takes the key findings from the first three sections and integrates them into a flow conceptualization for Yucca Flat.

Description of Study Area

Yucca Flat is in the northeastern part of the NNSS in Nye County, southern Nev., and is about 80 mi northwest of Las Vegas, Nev. (fig. 1). Most of the study area boundary coincides with the Yucca Flat hydrographic area (Cardinalli and others, 1968; Rush, 1968). The study area boundary differs only along its northwestern extent where the boundary is truncated along the eastern base of the Eleana Range.

Yucca Flat is a topographically closed drainage basin in the northern part of the Mojave Desert and ranges in altitude from about 3,920 ft in Yucca Lake to about 4,500 ft along the perimeters of the basin floor. Yucca Flat is bounded by low mountain ranges of Tertiary volcanic rocks and Paleozoic and Late Proterozoic sedimentary rocks (Slate and others, 1999) that typically reach altitudes of 5,000–7,000 ft. In general, the valley floor slopes upward toward the surrounding mountain ranges on a series of coalescing alluvial fans that ring the margins of the basin. Large, active alluvial channels extend into the basin from topographic highlands of Rainier Mesa and the Eleana and Halfpint Ranges (Slate and others, 1999; figs. 1, 2).

The arid climate of the study area is characterized by hot summers and mild to cold winters, large fluctuations in daily and annual temperatures, low precipitation and humidity, and occasional high winds. Average daily summertime maximum temperatures exceed 90°F and daily wintertime minimum temperatures average 20–25°F (Soulé, 2006). Annual precipitation ranges from about 6–7 inches on the valley floor to about 9–10 inches on the highlands bordering or just outside the study area (Blainey and others, 2007). Precipitation falls primarily as rain, but with occasional snow during the winter months at high altitudes. The limited runoff is conveyed from upland to lowland areas through washes that normally are dry. Shallow ponding occasionally occurs during spring in Yucca Lake playa at the southern end of Yucca Flat (fig. 2).

The NNSS was operated as the primary continental location for testing nuclear devices between 1951 and 1992. During its period of operation, 659 underground nuclear tests were conducted in Yucca Flat and 3 were conducted to the north of Yucca Flat near Climax Mine (U.S. Department of Energy, 2000a). Three of the underground tests in Yucca Flat are classified as crater tests, in which a nuclear device intentionally is buried shallow in order to eject the soil above the device during the explosion and produce a crater (U.S. Department of Energy, 2000b). All underground tests in the study area were conducted in alluvial and volcanic rock, and in a few cases in carbonate and granitic rock. The tests were emplaced in vertical shafts, or in a few cases, in mined drifts at the bottom of shafts. Many of the tests had small yields (less than 20 kilotons) and the largest test had an announced yield of 200 to 500 kilotons (U.S. Department of Energy, 2000a). Some of the tests consisted of multiple, simultaneous detonations completed in the same or adjacent boreholes (U.S. Department of Energy, 2000b). This resulted in a total of 750 underground nuclear detonations (including three crater detonations) in the study area (U.S. Department of Energy, 2000b). The surface of Yucca Flat is marked by hundreds of collapse sinks or subsidence craters that formed at the locations of underground nuclear detonations (Grasso, 2000, 2001; Stoller-Navarro Joint Venture, 2009). Some of these collapse sinks in the central part of Yucca Flat are evident on the shaded-relief base map shown on figure 2.

Some of the underground tests had large and sustained effects on water levels (Fenelon, 2005). Most of the effects are near-field, within a few thousand feet of a test. In some cases, tests in the volcanic confining unit elevated water levels more than 1,000 ft by pressurizing pore water within the low-permeability unit. Understanding differences between predevelopment and modern-day groundwater flow fields requires identifying the test-affected water levels.

Geologic and Hydrologic Setting

The geology of the Yucca Flat study area includes a locally thick Cenozoic volcanic and sedimentary section that unconformably overlies previously deformed rocks of Proterozoic through Paleozoic age (Winograd and Thordarson, 1975; Laczniak and others, 1996; Bechtel Nevada, 2006). The pre-Cenozoic section includes, from oldest to youngest: (1) greater than 9,000 ft of Late Proterozoic and Lower Cambrian siliceous and argillaceous metasediments (Stewart, 1970; 1972; Bechtel Nevada, 2006); (2) up to 15,000 ft of Middle Cambrian through Devonian dolomite, interbedded limestone, and thin but persistent shale and quartzite layers (Stewart and Poole, 1974; Poole and others, 1992; Bechtel Nevada, 2006); (3) as much as 8,000 ft of Mississippian siliceous siltstone, sandstone, and conglomerate (Poole and Sandberg, 1977; Trexler and others, 1996; Cole and Cashman, 1999; Bechtel Nevada, 2006); (4) a relatively thin Pennsylvanian limestone that locally overlies the Mississippian siliciclastic section; and (5) local granitic intrusive bodies.

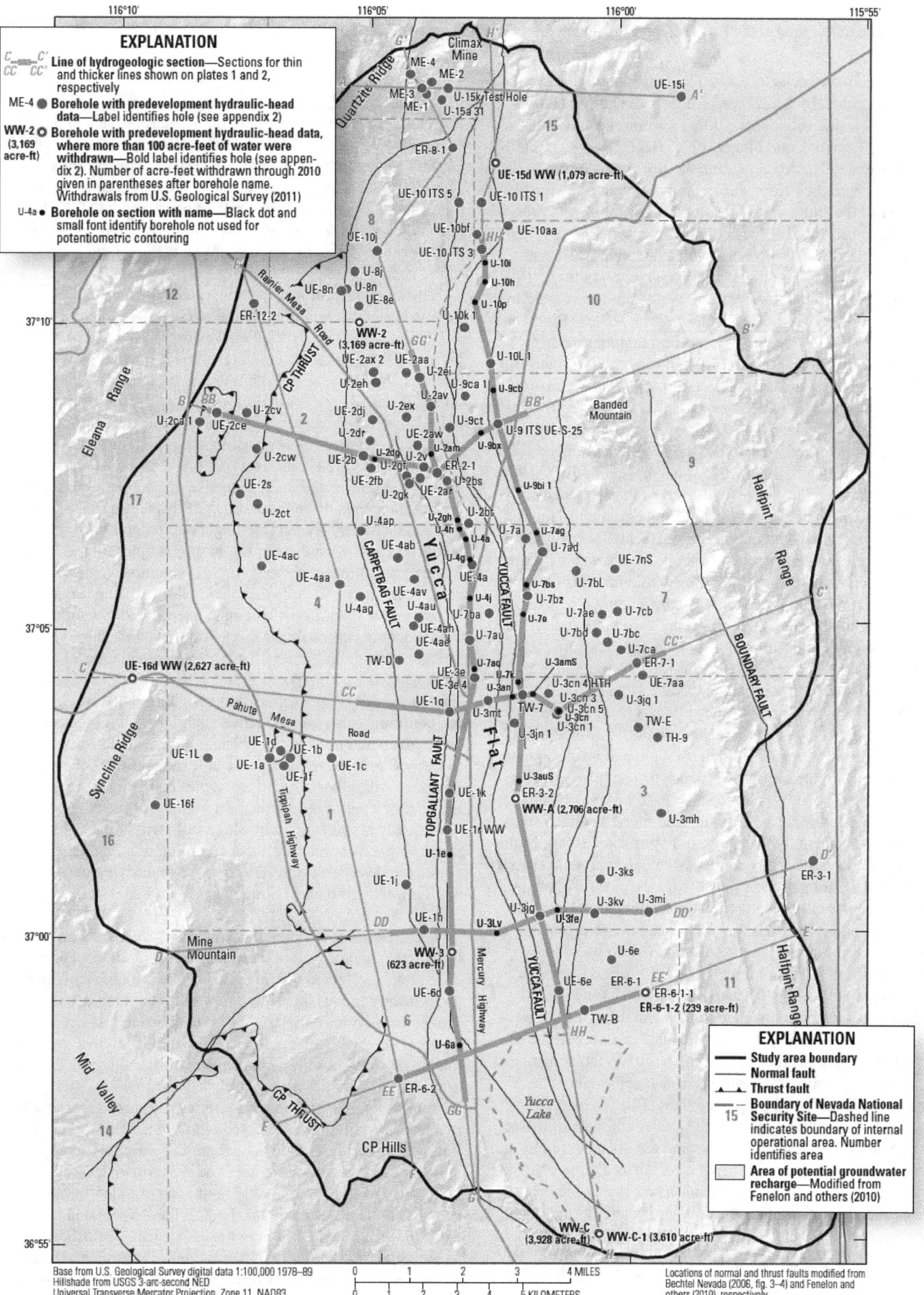

Figure 2. General features, major geologic structures, and location of boreholes with predevelopment hydraulic heads used to develop potentiometric contours in Yucca Flat, Nevada National Security Site.

The Cenozoic rocks in Yucca Flat consist of a thick alluvial section underlain by Miocene volcanic rocks and uncommon, thin, pre-volcanic sedimentary rocks (Winograd and Thordarson, 1975; Laczniak and others, 1996; Bechtel Nevada, 2006). The Miocene volcanic rocks form the lower part of the basin fill and were erupted from source areas located to the north and west of the study area. Many of the volcanic units present in the subsurface of Yucca Flat were erupted from vents within the southwestern Nevada volcanic field (Byers and others, 1976; Carr and others, 1986; Sawyer and others, 1994); however some of the oldest volcanic units have poorly defined source areas that may lie to the north of the NNSS (Ekren and others, 1971). The volcanic-rock section includes variably welded ash-flow tuff, ash-fall tuff, and reworked tuff and is up to 2,500 ft thick in the central part of the Yucca Flat basin (IT Corporation, 1996; Bechtel Nevada, 2006; Wood, 2007). The lower part of the volcanic-rock sequence primarily consists of bedded and reworked, nonwelded, commonly zeolitized tuffs (Winograd and Thordarson, 1975; Laczniak and others, 1996; Bechtel Nevada, 2006). The upper part of the sequence consists mainly of welded ash-flow tuff (Winograd and Thordarson, 1975; Laczniak and others, 1996; Bechtel Nevada, 2006). The post-volcanic sedimentary basin fill consists of a mixture of loosely consolidated coarse-grained alluvial and colluvial deposits that were derived from the surrounding Cenozoic volcanic and Paleozoic siliciclastic and carbonate sedimentary rocks, fine-grained basin-axis and playa deposits, and localized eolian sand and sporadic basalt flows (Winograd and Thordarson, 1975; Sweetkind and Drake, 2007b).

The region surrounding Yucca Flat has been affected by two opposing styles of tectonic deformation: mid-Mesozoic through Eocene compressive deformation, and a subsequent phase of mid-to-late Cenozoic extension. The pre-Cenozoic section in Yucca Flat was affected by east-west shortening in the form of regional thrust faults and more localized folds (Cole and Cashman, 1999; Snow and Wernicke, 2000). In the vicinity of Yucca Flat, this compressive deformational episode resulted in the formation of the CP thrust (fig. 2). This fault is locally exposed in outcrop in the southwestern part of the study area, and the fault is known to exist in the subsurface of western Yucca Flat from drill-hole data (Caskey and Schweickert, 1992; Cole and others, 1997; Cole and Cashman, 1999; National Security Technologies, 2008). Extensional Cenozoic deformation resulted in the formation of a series of north-striking, east-dipping, down-to-the-east normal faults within the Yucca Flat basin. The largest-offset faults displace the Cenozoic section by hundreds of feet and include the Yucca, Topgallant, and Carpetbag faults (fig. 2). Numerous east- and west-dipping faults with smaller amounts of offset are subsidiary to these main faults and are documented by subsurface drill-hole data (IT Corporation, 1996; Bechtel Nevada, 2006; Wood, 2007). Structural patterns change at the far northern and southern ends of the basin, where normal faults change orientation and sense of offset (Cole and Cashman, 1999; Hudson, 1992).

The subsurface extent of the geologic units in Yucca Flat have been defined through interpretation of drill-hole data from numerous wells, construction of interpretive geologic cross sections, and geophysical data (IT Corporation, 1996; Bechtel Nevada, 2006; Wood, 2007). These diverse data served as the primary input for the construction of a three-dimensional hydrostratigraphic framework model of the basin (Bechtel Nevada, 2006). The three-dimensional model serves as the primary basis for understanding the geologic framework of the basin.

The mapped geologic units in Yucca Flat have been grouped as hydrostratigraphic units (HSUs) on the basis of similar geologic and hydraulic properties (Winograd and Thordarson, 1975; Laczniak and others, 1996; Bechtel Nevada, 2006; National Security Technologies, 2007a; Fenelon and others, 2010). Three general groupings of the most permeable HSUs have been classified as aquifers: basin-fill alluvial deposits, volcanic rocks consisting of welded tuffs and lava flows, and fractured carbonate rocks. The principal carbonate aquifer consists of the thick sequence of Paleozoic carbonate rock that extends throughout much of the subsurface of central and southeastern Nevada (Dettinger and others, 1995; Harrill and Prudic, 1998) and crops out in the Halfpint Range in the eastern part of the study area (fig. 2). Fractured Cenozoic volcanic rock and permeable Cenozoic basin-fill alluvium form important local aquifers that contribute flow to the underlying Paleozoic carbonate aquifer (Winograd and Thordarson, 1975; Dettinger and others, 1995; Fenelon and others, 2010). Groundwater flow in Yucca Flat is obstructed or diverted by low-permeability rocks that form confining units. Rocks forming confining units include siliciclastic rock, granitic rock, and bedded and nonwelded volcanic tuffs (fig. 3). Proterozoic to Early Cambrian metamorphic and siliciclastic rocks and Paleozoic siliciclastic rock form a basement confining unit, which crops out in the northeastern corner of the study area. Zeolitically altered and nonwelded tuffs within the volcanic section form an important local confining unit that separates alluvial and volcanic aquifers from the underlying carbonate aquifer (Winograd and Thordarson, 1975; Fenelon and others, 2010).

Aquifers form regional or local flow systems depending on their extent and degree of interconnection. Widespread interconnected aquifers make up regional flow systems in which groundwater moves, nearly unimpeded, over long distances (Fenelon and others, 2010). Poorly connected, less extensive aquifers make up isolated to semi-isolated local flow systems that commonly provide a source of diffuse leakage or local drainage to an adjacent or underlying regional flow system. Diffuse leakage usually occurs over a widespread area at a low rate and most often is associated with limited flow across an intervening confining unit. Local drainage usually occurs at a higher rate over a limited area and most often is associated with flow through a permeable fault zone or along/through the zone of contact between a local aquifer and a regional aquifer.

Most groundwater flowing beneath Yucca Flat originates as precipitation falling in highland areas. In Yucca Flat, water recharges the groundwater flow system locally in highland areas such as the Eleana Range, Syncline Ridge, Mine Mountain, CP Hills, and the highest areas of the Halfpint Range and along the Belted Range (figs. 1, 2). Precipitation, and subsequent recharge, may be concentrated in ephemeral channels draining the highlands. Some recharge, specifically associated with water moving through the carbonate aquifer, is derived from precipitation falling on upland and mountainous areas north of the study area in central Nevada. Other potential pathways for small amounts of recharge are through fissures in Yucca Lake (Doty and Rush, 1985) and through subsidence craters created by underground nuclear detonations (Hokett and others, 2000), which focus and capture nearby surface runoff.

Recharge occurs when precipitation falling on highland areas collects in surface fractures and openings and infiltrates downward by way of interconnected fractures or through the rock matrix to depths beyond the influence of active evaporation and transpiration. The presence of less-permeable rock can impede the downward movement of water, thereby creating zones of perched or semi-perched groundwater (Winograd and Thordarson, 1975). The term "semi-perched" is used to differentiate zones of shallow, elevated water that are underlain by saturated rocks; perched zones, by definition, are underlain by unsaturated rocks (Meinzer, 1923).

Water within unsaturated rock or in perched or semi-perched zones beneath underground test areas in Yucca Flat may move radionuclides downward into saturated, permeable rock. Once within saturated rock, transport of the radionuclides is controlled in part by the rate and direction of groundwater flow, which itself is controlled by the permeability of the host rock and by differences in hydraulic head (or hydraulic gradient).

Groundwater that reaches the regional flow system generally flows toward discharge areas south and southwest of Yucca Flat (fig. 1; Fenelon and others, 2010). Here groundwater discharges from springs or by diffuse upward flow into an overlying shallow flow system, where the water is evaporated, or transpired by phreatophytes. Major areas discharging groundwater flowing through Yucca Flat likely include Ash Meadows (fig. 1), Franklin Lake, and Death Valley further southwest (Winograd and Thordarson, 1975; Laczniak and others, 1996; Fenelon and others, 2010). A discharge mechanism locally important within Yucca Flat is the removal of groundwater by pumping. Since 1952, groundwater has been withdrawn from wells open to primarily alluvium and carbonate rock throughout Yucca Flat (fig. 2).

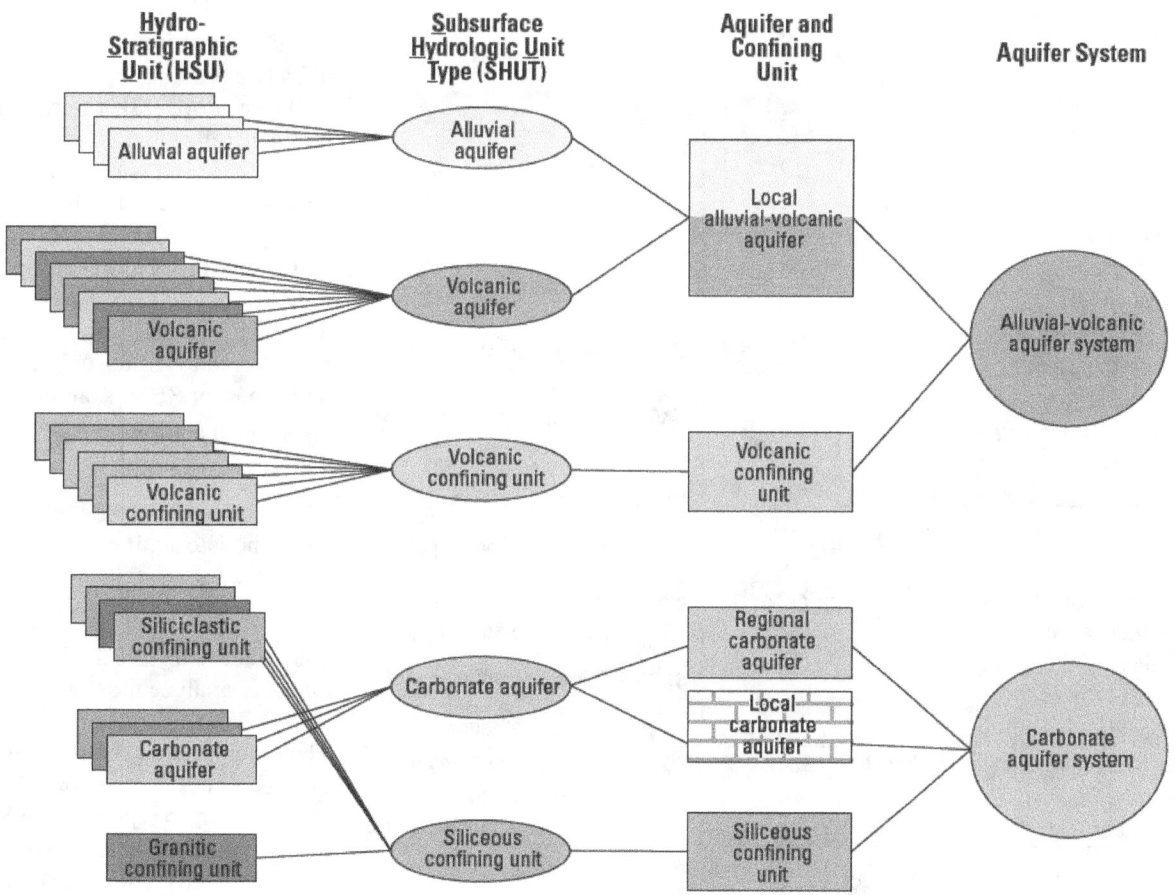

Figure 3. Aquifer and confining unit classification system used to conceptualize groundwater flow in Yucca Flat, Nevada.

Study Methods

The general approach used to develop a conceptualization of groundwater flow through the study area was to delineate the extent of the permeable rocks forming the three primary aquifer types—alluvial, volcanic, and carbonate—and the less permeable rocks forming the two primary confining unit types—volcanic and siliceous (fig. 3). Discrete aquifers identified within each of these aquifer types are classified as either regional or local. Regional aquifers typically include large, spatially extensive blocks of permeable rock that together form part of a larger flow system, whereas local aquifers typically are areally less extensive, hydraulically isolated, and drain only to adjacent confining units. Hydraulic heads in the principal aquifers and the volcanic confining unit are contoured to determine general flow directions and the interconnection between aquifers.

Delineation of Aquifer Systems

An aquifer system, for the purpose of this report, is defined as a grouping of aquifers and confining units that generally function together as an individual flow system. The confining units that are part of an aquifer system generally bound the aquifers within the system, preventing significant flow interaction with adjacent aquifer systems. Two aquifer systems were identified in the study area: alluvial–volcanic and carbonate (fig. 3).

The first step in the flow conceptualization process was to identify and delineate the aquifer systems and their component aquifers and confining units in Yucca Flat. These aquifers and confining units were identified and mapped using a composite hydrostratigraphic framework developed by merging two three-dimensional hydrostratigraphic framework models (HFMs) constructed previously as part of the U.S. Department of Energy's Underground Test Area Project (U.S. Department of Energy, 2003). The models support investigations of radionuclide groundwater contamination in the Yucca Flat–Climax Mine and Rainier Mesa–Shoshone Mountain areas (Bechtel Nevada, 2006; National Security Technologies, 2007a).

Each HFM is composed of hydrostratigraphic units (HSUs) that consist of one or more stratigraphic units with similar geologic and hydraulic properties. Twenty-six HSUs extending below the water table in the study area were identified in the two HFMs (fig. 4). The saturated parts of these HSUs form the hydrogeologic foundation used to develop the conceptualization of groundwater flow presented in this report. The HSUs evaluated as part of this study include 15 aquifers and 11 confining units (fig. 4).

Framework HSUs were grouped into generalized unit types on the basis of rock type and whether the HSU was classified as an aquifer or confining unit (fig. 3). Combining HSUs of similar hydrologic and geologic properties reduced the number of subsurface units to five (figs. 3, 4). These five units herein are referred to as subsurface hydrologic unit types, or SHUTs, and include:

- alluvial aquifer
- volcanic aquifer
- volcanic confining unit
- carbonate aquifer
- siliceous confining unit

The SHUTs presented here are consistent with SHUTs developed to map and analyze the flow systems at the NNSS (Fenelon and others, 2010). The color scheme shown on figure 3 is used on plates 1–4 and on other figures in this report to represent the SHUTs: alluvial aquifer (yellow); volcanic aquifer (green); volcanic confining unit (orange); carbonate aquifer (blue); siliceous confining unit (brown).

HSU name	HSU	SHUT	SHUT name
Alluvial aquifer	AA		
Alluvial aquifer 3	AA3	AAQ	Alluvial aquifer
Alluvial aquifer 2	AA2		
Alluvial aquifer 1	AA1		
Timber Mountain upper vitric-tuff aquifer	TMUVTA		
Timber Mountain welded-tuff aquifer	TMWTA	VAQ	Volcanic aquifer
Timber Mountain lower vitric-tuff aquifer	TMLVTA		
Upper tuff confining unit	UTCU	VCU	Volcanic confining unit
Topopah Spring aquifer	TSA	VAQ	Volcanic aquifer
Belted Range aquifer	BRA		
Belted Range confining unit	BRCU	VCU	Volcanic confining unit
Lower tuff confining unit	LTCU		
Tub Spring aquifer	TUBA	VAQ	Volcanic aquifer
Pre-Grouse Canyon tuff lava flow aquifer 1	PRETBG_1		
Oak Spring Butte confining unit	OSBCU	VCU	Volcanic confining unit
Redrock Valley welded-tuff aquifer	RVA	VAQ	Volcanic aquifer
Lower tuff confining unit 1	LTCU1	VCU	Volcanic confining unit
Argillic tuff confining unit	ATCU		
Mesozoic granite confining unit	MGCU	SCU	Siliceous confining unit
Upper carbonate aquifer	UCA	CAQ	Carbonate aquifer
Upper clastic confining unit 1	UCCU1	SCU	Siliceous confining unit
Upper clastic confining unit	UCCU		
Lower carbonate aquifer thrust plate	LCA3	CAQ	Carbonate aquifer
Lower clastic confining unit thrust plate 2	LCCU2	SCU	Siliceous confining unit
Lower carbonate aquifer	LCA	CAQ	Carbonate aquifer
Lower clastic confining unit	LCCU	SCU	Siliceous confining unit

For more information on these units, see Microsoft® Excel worksheet "SHUTtoHSU_Chart" in appendix 3

Figure 4.   Correlation of hydrostratigraphic units (HSUs) and subsurface hydrologic unit types (SHUTs) occurring in the saturated zone in the Yucca Flat study area, Nevada.

The configuration and extent of the saturated parts of each SHUT were extracted from the HFMs and mapped for use in potentiometric contouring. In the area of overlap between the two HFMs, the more current Rainier Mesa–Shoshone Mountain HFM took precedence. The two-dimensional mapping was combined with hydrogeologic cross sections (pls. 1 and 2) developed from vertical slices of the HFMs to develop a three-dimensional configuration of the aquifers and confining units for this study. The mapped aquifers and confining units were grouped into two aquifer systems that defined two semi-independent flow systems (fig. 3).

Analysis of Water Levels

Water levels from 229 completion zones or wells (app. 1) in 166 boreholes were compiled, reviewed, and analyzed. Many of the boreholes are concentrated in areas of past underground testing in the central part of Yucca Flat. As used in this report, a well is defined as a single, temporary or permanent completion in a borehole, where each completion defines a unique set of open intervals. By this definition, many boreholes in the study area contain multi-well completions. Multi-well boreholes may consist of temporary completions where measurements are made in packed-off intervals or permanent completions, such as multiple monitoring tubes installed within the annulus of a main well completion. Naming conventions for the wells and boreholes referred to in this report are as follows. A well that is the sole completion interval in a borehole is assigned the name of the borehole. In boreholes with multiple completions, wells are differentiated by names that use a parenthetical expression added after the borehole name—for example: *UE-12t 6 (1378 ft)*. A single number in the parenthetical expression refers to the depth of the well; two numbers separated by a dash refer to the depth of the top and bottom of the open interval in the well. In some cases, a well name consists of the borehole name followed by one of three non-parenthetical expressions: main, piezometer, or WW. All well names in this report are consistent with those used in the U.S. Geological Survey National Water Information System (NWIS) database and are italicized in the text for clarity. Well names, borehole names as used in this report, and official NNSS borehole names are cross-referenced in appendix 2.

Approximately 5,200 water levels in 229 wells were measured in the Yucca Flat study area from 1951 to 2010 (app. 1). The depth to water exceeds 1,000 ft in the majority of the wells. These water levels were used to determine predevelopment hydraulic-head distributions in the groundwater system. Each water-level measurement was reviewed for correctness and accuracy, assigned to an open interval, examined to determine the hydrologic condition at the time of measurement, and flagged to indicate if the level reflects predevelopment hydrologic conditions or transient conditions imposed by nuclear testing or pumping. The thorough evaluation ensures data integrity and identifies water levels that best represent predevelopment conditions for hydraulic-head contouring. A large part of the water-level analysis was supported by comprehensive evaluations of water levels in Yucca Flat and the NNSS area (Fenelon, 2005; Elliott and Fenelon, 2010). Water levels and well-construction information are stored in the USGS National Water Information System (NWIS) database and can be accessed from the internet at *http://waterdata.usgs.gov/nv/nwis/gw*.

As part of the water-level analysis, each water level in appendix 1 was flagged to indicate whether it is representative of each of the following three hydrologic conditions: (1) natural predevelopment conditions, (2) transient conditions resulting from nearby nuclear testing, and (3) transient conditions resulting from pumping. Assignment of a predevelopment-condition flag assumes that human activity has not affected or has minimally affected the water level. For example, a recent water-level measurement that is believed to be influenced primarily by natural climatic fluctuations is considered representative of predevelopment conditions. Determining whether a water level represents predevelopment or human-induced transient conditions sometimes was difficult. Difficulties arose primarily in wells with few water-level measurements or in wells open to confining units, where vertical hydraulic gradients are large and water levels are naturally elevated and equilibrate slowly. In these situations, determining whether an elevated water level is equilibrated and representative of natural conditions or affected by nuclear testing often is problematic.

Several factors were used to determine whether a water level in a confining unit was representative of predevelopment conditions. Water levels from a well in a confining unit typically equilibrate slowly from stresses to the system, such as a nearby nuclear test or drilling and developing the well. Water levels from confining units in Yucca Flat that are representative of predevelopment conditions typically change little from year to year. If sufficient measurements are available, trending water levels suggest nonequilibrated conditions and stable water levels suggest equilibrated conditions. In wells where no clear trend could be determined because water-level data were sparse or of short measurement duration, other factors were considered. These include the elapsed time between the measurement and well completion, the consistency of the measured water-level altitude relative to nearby water-level altitudes, the length of open interval at the well, and any knowledge about the hydraulic conductivity of the confining unit materials open to the well.

Because of the difficulty in determining whether a water level represents one or more of the three hydrologic conditions (predevelopment, nuclear testing, and pumping), each water level is assigned one of the following five uncertainty flags for each hydrologic condition:

- "Yes"—Water level represents hydrologic condition.
- "Yes?"—Water level probably represents hydrologic condition, but assignment is uncertain.
- "?"—Water level may or may not represent hydrologic condition.
- "No?"—Water level probably does not represent hydrologic condition, but assignment is uncertain.
- "No"—Water level does not represent hydrologic condition.

The assignment of these five uncertainty flags allowed for qualitative weighting of the water levels in later analyses. For example, during the contouring process, more weight was given to a predevelopment water level assigned an uncertainty flag of "Yes" than to a water level assigned a flag of "Yes?" or "?".

Well hydrographs, well locations, water levels, and flag assignments can be displayed interactively from a Microsoft® Excel workbook (app. 1). The workbook is designed to be an easy-to-use tool to view water levels and other information associated with wells in the study area. Information for an individual well can be selected by using the AutoFilter option available in Excel. An example of the information available for two wells in the appendix is provided on figure 5.

Hydraulic head at each well opening is equated to the water-level altitude in the well. However, hydraulic head depends on the density of the water, which in the study area can vary due to differences in water temperature. Wells in the study area that have a long (several thousand feet) water column (app. 2) in combination with a warm water-column temperature (more than 10°F greater than typical groundwater temperatures in Yucca Flat of about 86°F) could have a temperature-equivalent hydraulic head that is several feet or more lower than would be computed directly from the depth-to-water measurement.

Temperature adjustments were not applied to hydraulic heads because these adjustments are considered minimal relative to horizontal hydraulic gradients in the study area. An attempt was made by Fenelon and others (2010) to adjust water-level measurements for variations in water temperature, primarily in order to account for potentially large (greater than 5 ft) errors that might mask or alter the true hydraulic gradient in areas of small horizontal or vertical head change. Fenelon and others (2010) computed temperature adjustments for the 16 wells in Yucca Flat that had more than 1,000 ft of water column above the mid-point of the open interval (app. 2); wells with shorter water columns were assumed to have small (less than 5 ft) temperature adjustments. Of the 16 wells analyzed, *ER-12-2 main (lower zone)* and *U-3cn 5* had temperature adjustments that exceeded 5 ft (17 and 8 ft, respectively). Hydraulic heads in these two wells were not adjusted in this report prior to water-level contouring because of large uncertainties in the adjustment. Uncertainties result from poorly constrained estimates of the parameters used to calculate the temperature adjustment, including average water-column temperature and zones of inflow into the well (Fenelon and others, 2010). The uncertainties in the hydraulic-head estimates in these two wells do not affect the conceptualization of groundwater flow in Yucca Flat.

Most hydraulic heads (water-level altitudes) computed from depth-to-water measurements given in appendix 1 are considered accurate to within 5 ft. In most cases, actual depth-to-water measurements made in Yucca Flat are accurate to 1 ft or less, depending on the method of measurement. Errors caused by borehole deviation in the conversion from depth-to-water to hydraulic head generally are less than 0.5 ft. Where errors are known to be larger, the measured water levels were corrected for borehole deviation (Elliott and Fenelon, 2010). Hydraulic heads for nine non-surveyed wells may be in error by 2 to 10 ft due to inaccuracies in estimates of land-surface altitude. Land-surface altitudes of all other wells are considered accurate to within 1 ft. The reported accuracy of the land-surface altitude for each well in the study area is provided in appendix 2.

Estimation of Predevelopment Hydraulic Heads

Water levels in each well were evaluated, as discussed in the previous section, to determine if and which water levels represent predevelopment hydrologic conditions. Of the 229 wells analyzed for this study, 166 of the wells (app. 2) had at least one water level identified as representative or potentially representative of predevelopment conditions. Locations and borehole names for these 166 wells are shown on figure 2.

A single estimate of hydraulic head was used to represent predevelopment conditions in each of the 166 wells identified as having at least one predevelopment water level. These wells are identified in appendix 2 and are those in which the dataset field "*Predevelopment map use of hydraulic head*" does not equal "*None*." For wells with multiple predevelopment measurements, the mean of the measurements was used as the predevelopment hydraulic-head estimate. Water levels used to estimate the predevelopment head at each of the 166 wells with at least one predevelopment water level are shown as large blue diamonds on hydrographs that can be plotted interactively by using appendix 1 (fig. 5). A synoptic set of water-level measurements for all wells in the study area is preferred to using mean water levels, but such a set could not be developed because many of the wells previously measured have been destroyed and current hydrologic conditions monitored by some existing wells no longer represent predevelopment conditions. The error associated with comparing water levels that span decades is assumed to be relatively minor because long-term, naturally occurring, water-level fluctuations in areas of low recharge such as Yucca Flat generally are less than 5 ft (Elliott and Fenelon, 2010).

The predevelopment estimate of the hydraulic head was determined from a single water-level measurement in 75 of the 166 wells. In more than one-half of these 75 wells, the single measurement could be used only as an upper or lower bound for the predevelopment head. For example, on a rising water-level hydrograph that is equilibrating toward predevelopment conditions, the last water level can be used as a lower bound for the expected predevelopment head in the well. In this example, if the altitude of the last water-level measurement was 1,000 ft, the predevelopment head is expected to be greater than 1,000 ft. For measurements made in a dry well, the altitude of the bottom of the well is assigned a "less than" qualifier and is used as an upper bound for contouring. Only hydraulic heads calculated from mean water levels representing predevelopment conditions, or those that were assigned a

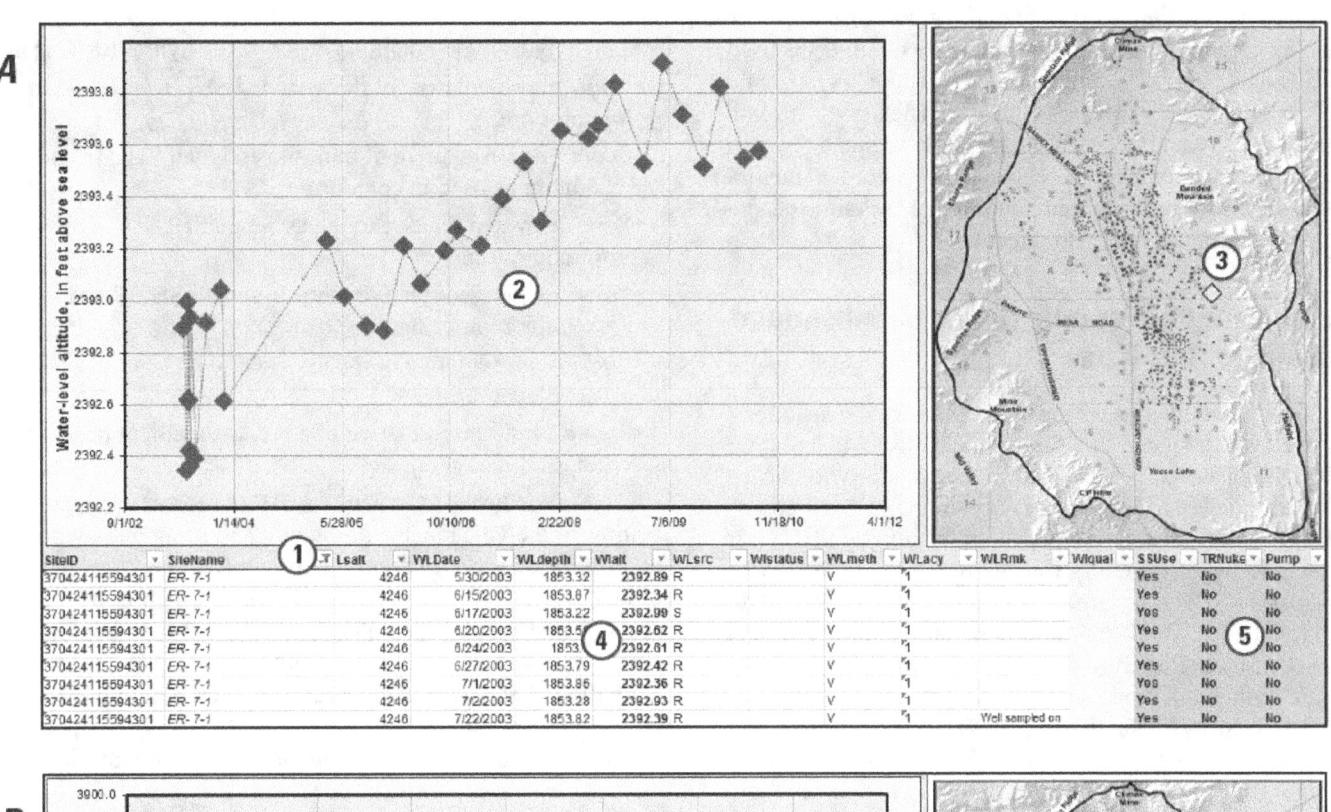

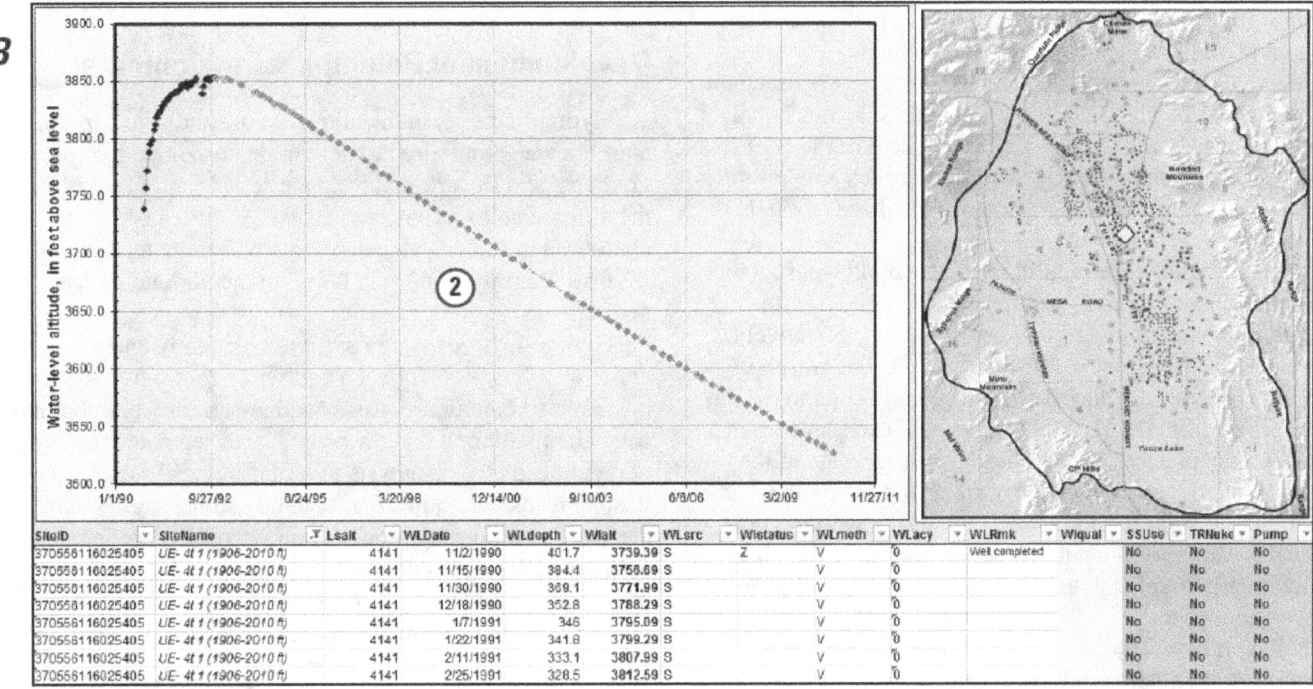

Plots highlight the type of information in the spreadsheet: (1) pull-down menu to select well of interest; (2) hydrograph of all water-level measurements for the selected well—measurements used in contouring are shown as blue diamonds, nonstatic measurements are shown in black, and measurements affected by nuclear testing are shown in red; (3) map showing the selected well location as a yellow symbol; (4) water-level data for the well, and (5) water-level flags indicating the likelihood that each water level represents predevelopment conditions, transient conditions resulting from nearby nuclear testing, or transient conditions from nearby pumping.

Figure 5. Examples from appendix 1 Microsoft® Excel workbook showing water levels analyzed in wells **A**, *ER-7-1* and **B**, *UE-4t 1 (1906–2010 ft)*.

qualifier to constrain the predevelopment head, were used to guide the contouring process. One of three qualifiers is used to describe a bounding hydraulic head: "less than" ($<$), "greater than" ($>$), or "greater than or equal" ($\geq$). The use of "$\geq$" indicates that the hydraulic head is most likely within a few feet of approximating a predevelopment head. A "$<$" or "$>$" qualifier provides no information about how close a hydraulic head is to approximating a predevelopment head.

Assignment of Hydraulic Heads to Subsurface Hydrologic Unit Types

The predevelopment estimate of the hydraulic head for each well was assigned to a subsurface hydrologic unit type (SHUT). The assignment is made in accordance with the SHUT encountered at the open interval (app. 2, 3). Wells with long open intervals commonly penetrate multiple SHUTs. In these cases, heads generally were associated with the most transmissive SHUT. In most cases, the top and bottom SHUT altitudes at each well location were determined from altitudes of HSU surfaces that were developed for the hydrostratigraphic framework models (HFMs; Bechtel Nevada, 2006; National Security Technologies, 2007a). In a few select cases, HSU surface altitudes were derived from an electronic database that supersedes the surface altitudes in the HFM reports (Sigmund Drellack, National Security Technologies, written commun., 2008). The altitudes of HSU surfaces are based on, and in good agreement with, lithologic picks from original lithologic logs (Wood, 2007). The assignment of the contributing SHUT based on the HSUs identified in each well provides a consistent method for assigning hydraulic heads to SHUTs across the entire study area.

The HSUs and corresponding SHUTs for all wells analyzed in this study (app. 2) and for all boreholes used for underground nuclear testing can be displayed interactively from a Microsoft© Excel workbook (app. 3). The workbook is designed to view (1) the hydrostratigraphic column, which is interpreted from the HFM, (2) the mean water level used to develop the predevelopment head, or an estimate of the predevelopment water level, (3) basic well-construction information for wells in the study area, and (4) the working point (vertical location) of the nuclear detonation, where applicable. The hydrostratigraphic columns provided in appendix 3 extend to the bottom of the HFM, about 16,000 ft below sea level. HSUs shown below the bottom of a borehole are derived from the HFM and are model interpretations. In areas where the two HFMs overlap, the more recently developed Rainier Mesa–Shoshone Mountain HFM was used to assign a HSU. Information for an individual well can be viewed in appendix 3 by selecting the well or borehole from the column-header dropdown list. Two examples from the workbook page, one showing an emplacement borehole for a nuclear detonation and the other showing a monitoring well, are presented on figure 6.

Each hydraulic-head estimate is assigned a single hydrologic qualifier that describes how the estimate was used in the process of contouring hydraulic heads (app. 2). The six assigned hydrologic qualifiers describe the hydraulic head as:

- "*rep*", representative of the assigned SHUT and used in contouring.
- "*com*", representative of multiple assigned SHUTs and sometimes used in contouring.
- "*elv*", elevated relative to the assigned SHUT and not used for contouring.
- "*low*", depressed relative to the assigned SHUT and not used for contouring.
- "*lim*", representative of the assigned SHUT but of limited use in contouring.
- "*none*", not representative of a predevelopment head and not used in contouring.

Hydraulic heads assigned to a SHUT representing an aquifer or the volcanic confining unit and designated with a hydrologic qualifier of "rep", "com", "elv", and "low" were plotted on plates 3 and 4. All heads with a qualifier of "rep" and selected heads with a qualifier of "com" were contoured. Heads assigned to the siliceous confining unit SHUT (app. 2) were plotted and used to constrain contours within the aquifers. Heads with hydrologic qualifiers of "lim" and "none" are not shown on the plates and were not used for contouring.

Development of Potentiometric Contours

Hydraulic heads in the alluvial, volcanic, and carbonate aquifers were contoured to determine horizontal hydraulic gradients and flow directions. One set of contours was used for the alluvial and volcanic aquifers. Hydraulic heads also were contoured in the volcanic confining unit to better conceptualize flow from the confining unit into the alluvial–volcanic aquifer.

Hydraulic heads were contoured manually and are posted at the well locations for each aquifer or confining unit. Each posted head contains an associated uncertainty flag that indicates the likelihood that the posted head represents a predevelopment head (see "Analysis of Water Levels" section above). Posted heads that indicate uncertainty sometimes were ignored during the contouring process, and, therefore, are inconsistent with potentiometric contours on plates 3 and 4. In most cases, the inconsistency between posted and contoured heads can be attributed to heads that likely are erroneously identified as representative of predevelopment conditions, large local vertical hydraulic gradients, unrecognized hydrologic anomalies, or measurement errors. Posted hydraulic heads disregarded during contouring are shown in italics on plates 3 and 4.

As part of the manual contouring process, potentiometric contours were configured in accordance with known or inferred hydraulic gradients, recharge areas, discharge areas, lateral and vertical continuity of flow systems, and the known or inferred geology. Specific examples of this manual process include the following:

- In areas where a fault juxtaposes a confining unit to form the boundary of an aquifer, contours are constructed perpendicular or nearly perpendicular to the inferred flow barrier (see 2,460-ft contour on fault-bounded block of alluvial – volcanic aquifer in western part of NNSS Area 9 on pl. 3).

- In areas where a fault or fault zone is inferred to impede flow within the aquifer, contours are configured in a tighter pattern to portray an increase in the local head

gradient upgradient of the inferred flow barrier (see 2,410-ft through 2,440-ft contours west of Yucca fault in western NNSS Area 3 on pl. 3).

- In areas where a fault or fault zone is inferred to be highly transmissive, contours are constructed convexly to the general flow direction to reflect a preferred flow path (see 2,400-ft contour on pl. 4).

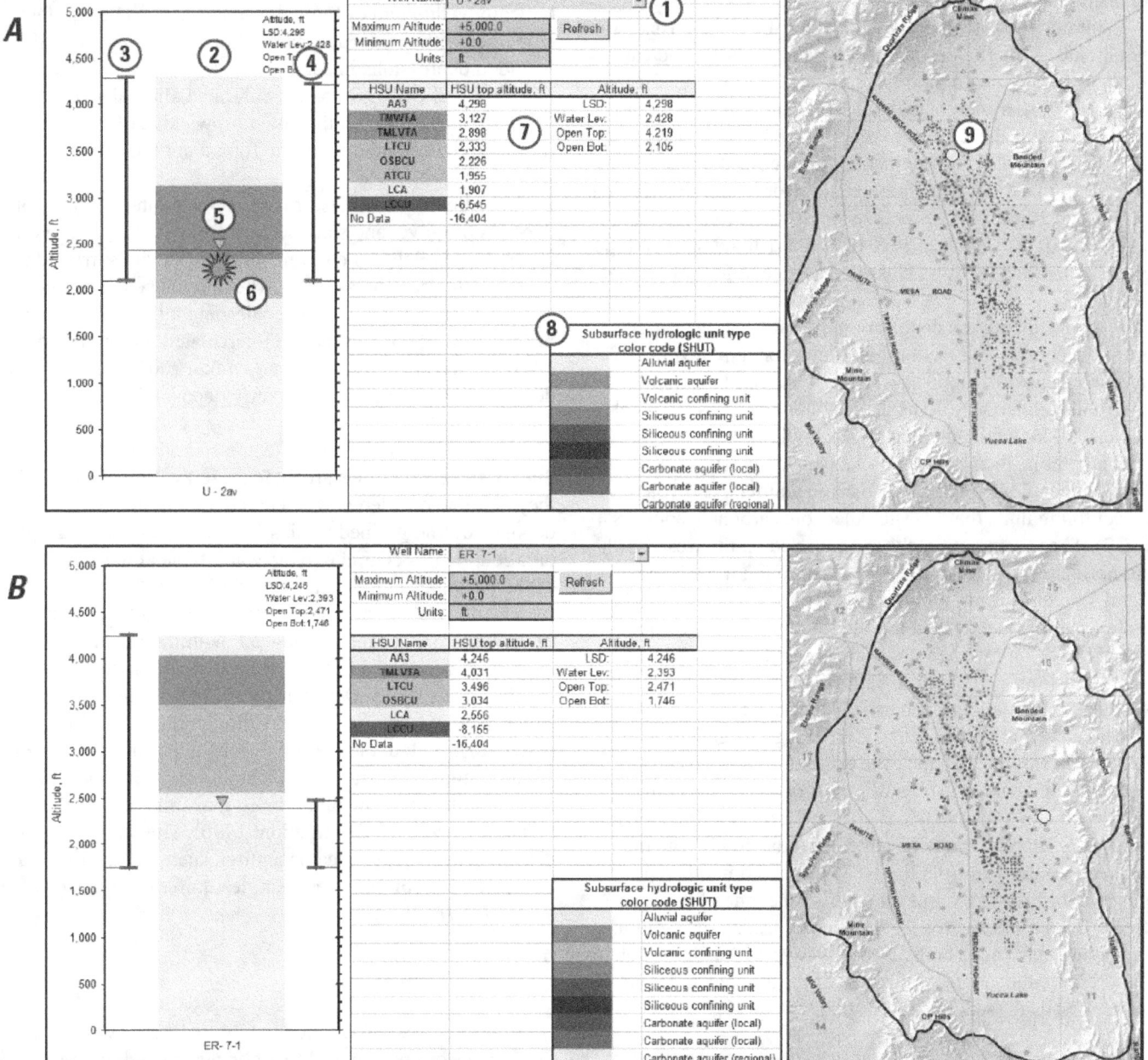

These are 2 of 941 Yucca Flat sites (including all underground nuclear detonations) in the workbook that show the hydrostratigraphic units penetrated by each borehole and their relation to water level and open intervals. Top plot highlights the type of information in the spreadsheet: (1) pull-down menu to select well or borehole of interest; (2) hydrostratigraphic column for selected well; (3) vertical extent of drilled borehole; (4) uppermost and lowermost extent of open interval which, if saturated, could contribute water to well; (5) measured or estimated water level; (6) nuclear detonation working-point depth; (7) tabular information specific to selected well or borehole; (8) explanation for color scheme of hydrostratigraphic column; and (9) map of the Yucca Flat study area showing selected well location as a yellow symbol.

Figure 6. Examples from appendix 3 Microsoft® Excel workbook showing *A*, borehole *U-2av*, which was used for emplacement of the Calabash nuclear detonation, and *B*, well *ER-7-1*, which was completed for hydrogeologic investigations.

Hydrogeologic Framework

The hydrogeologic framework in Yucca Flat is made up of multiple aquifers and confining units and the faults that dissect them. This framework has significant influence on local groundwater flow in Yucca Flat. As discussed, the framework defines the major aquifers and confining units, their lateral and vertical extents, their physical characteristics, and the interconnections between aquifers.

Saturated hydrostratigraphic units (HSUs) in the Yucca Flat area include Proterozoic and Paleozoic siliciclastic and carbonate rocks, Mesozoic intrusive rocks, Tertiary-age volcanic rocks, and Tertiary- and Quaternary-age alluvium (Winograd and Thordarson, 1975; Laczniak and others, 1996; Bechtel Nevada, 2006). HSUs were identified in these studies on the basis of similarity in stratigraphic and hydraulic properties and were grouped into five generalized subsurface hydrologic unit types, or SHUTs (fig. 3). The SHUTs are defined by rock type and whether the HSU was classified as an aquifer or confining unit (fig. 3).

A map showing the distribution of SHUTs at the water table was constructed from three-dimensional hydrostratigraphic framework models (HFMs) of Yucca Flat–Climax Mine and Rainier Mesa–Shoshone Mountain areas (Bechtel Nevada, 2006; National Security Technologies, 2007a). At the water table, the saturated rocks beneath most of Yucca Flat include alluvial aquifers, volcanic aquifers, and volcanic confining units that overlie Paleozoic carbonate aquifers (fig. 7). These saturated rocks are bounded on the west, north, and northeast by saturated siliceous confining units consisting of siliciclastic and granitic rocks (fig. 7). The subsurface configuration and extent of the saturated parts of each SHUT are portrayed on hydrogeologic cross sections (pls. 1 and 2) developed from vertical profiles from the HFMs (Bechtel Nevada, 2006; National Security Technologies, 2007a).

Large normal faults such as Carpetbag, Topgallant, and Yucca faults are located near the center of the basin (fig. 7). The three-dimensional HFM of Yucca Flat portrays both Topgallant and Yucca faults as consisting of a pair of subparallel, steeply-dipping faults (Bechtel Nevada, 2006). The maps and sections derived from the HFM presented in this report preserve this geometry, but often only label one of the fault strands for simplicity (fig. 7). The north-striking, east-dipping Topgallant and Yucca faults create two west-tilted half-grabens that preserve the greatest thickness of the alluvial and volcanic-rock SHUTs (sections C–C' and D–D', pl. 1).

Subsurface Hydrologic Unit Types

The five SHUTs mapped in Yucca Flat consist of three aquifers and two confining units. These include the alluvial, volcanic, and carbonate aquifers and the volcanic and siliceous confining units. Together, these SHUTs form the framework for the aquifer systems in Yucca Flat.

Alluvial Aquifer

Alluvial units are widespread in map view and occur as unsaturated deposits at land surface throughout most of the study area (fig. 8; Slate and others, 1999). The alluvial fill of the Yucca Flat basin ranges in thickness from a thin veneer along the margins of the valley to over 3,000 ft in south-central Yucca Flat (Bechtel Nevada, 2006). However, the saturated part of these young deposits that form the alluvial aquifer typically is thin and of limited extent (fig. 8). The aquifer coincides with the deepest parts of the post-volcanic fault-bounded depressions in southern Yucca Flat (fig. 8; sections DD–DD' and GG–GG', pl. 2). Saturated alluvial sediments are thicker than 1,000 ft on the downthrown side of Yucca fault (fig. 8) and the southern part of Topgallant fault (fig. 8; section GG–GG', pl. 2).

The alluvium consists of variably-cemented, poorly sorted deposits of gravel and sand derived from exposures of volcanic rocks and Paleozoic sedimentary rocks that surround the basin (Slate and others, 1999). The alluvial sediments were deposited in channels and on coalescing alluvial fans; deposits of fine-grained eolian sand are intercalated within the coarser alluvial deposits (Sweetkind and Drake, 2007b). The alluvial deposits range in age from recently deposited alluvium to the oldest tuff-bearing gravels that may correlate to the 9 Ma Thirsty Canyon Group in age (Slate and others, 1999). Local, thin, basaltic lava flows are interspersed within alluvium, but they are not considered as a separate unit in this report. Thick deposits of fine-grained sand, silt, and clay deposited as playa-lake sediments occur above the water table at the southern end of the basin beneath Yucca Lake (fig. 8).

The alluvial aquifer is important in Yucca Flat because of the underground tests conducted within it and the many more tests conducted in the unsaturated alluvium overlying the aquifer (U.S. Department of Energy, 2000b; Stoller-Navarro Joint Venture, 2009). Although relatively thick and widely distributed, the majority of the alluvium is unsaturated. Where saturated, the alluvium's high interstitial porosity and permeability allow the alluvium to transmit water efficiently (Claassen, 1973; IT Corporation, 1996). The alluvial aquifer directly overlies the volcanic aquifer, such that the two may be reasonably combined as a single aquifer (sections GG–GG' and HH–HH', pl. 2).

Volcanic Aquifer

Regionally extensive moderately to densely welded ash-flow sheets of the Timber Mountain Group and, in the southern half of Yucca Flat, the Paintbrush Group, form important volcanic aquifers in Yucca Flat (Winograd and Thordarson, 1975; Laczniak and others, 1996; Bechtel Nevada, 2006). Welded outflow-tuff sheets, which compose many of the volcanic aquifers at the NNSS, typically have well-connected fracture networks and minimal secondary alteration (Blankennagel and Weir, 1973). Fractured rhyolite lava flows and vitric ash-fall tuffs also are included as part of the volcanic

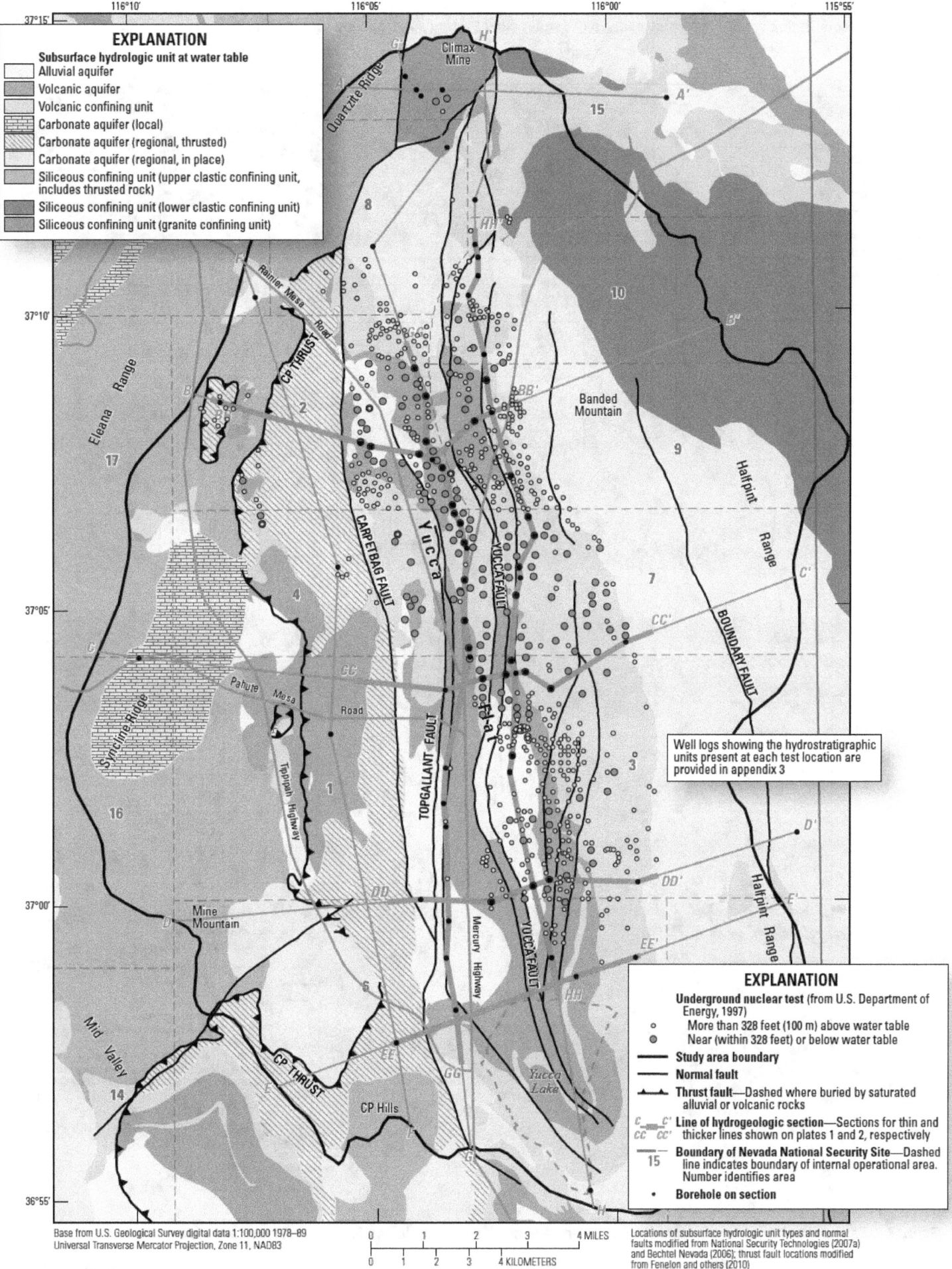

Figure 7. Distribution of subsurface hydrologic unit types at the water table and location of underground nuclear tests relative to the water table in Yucca Flat, Nevada.

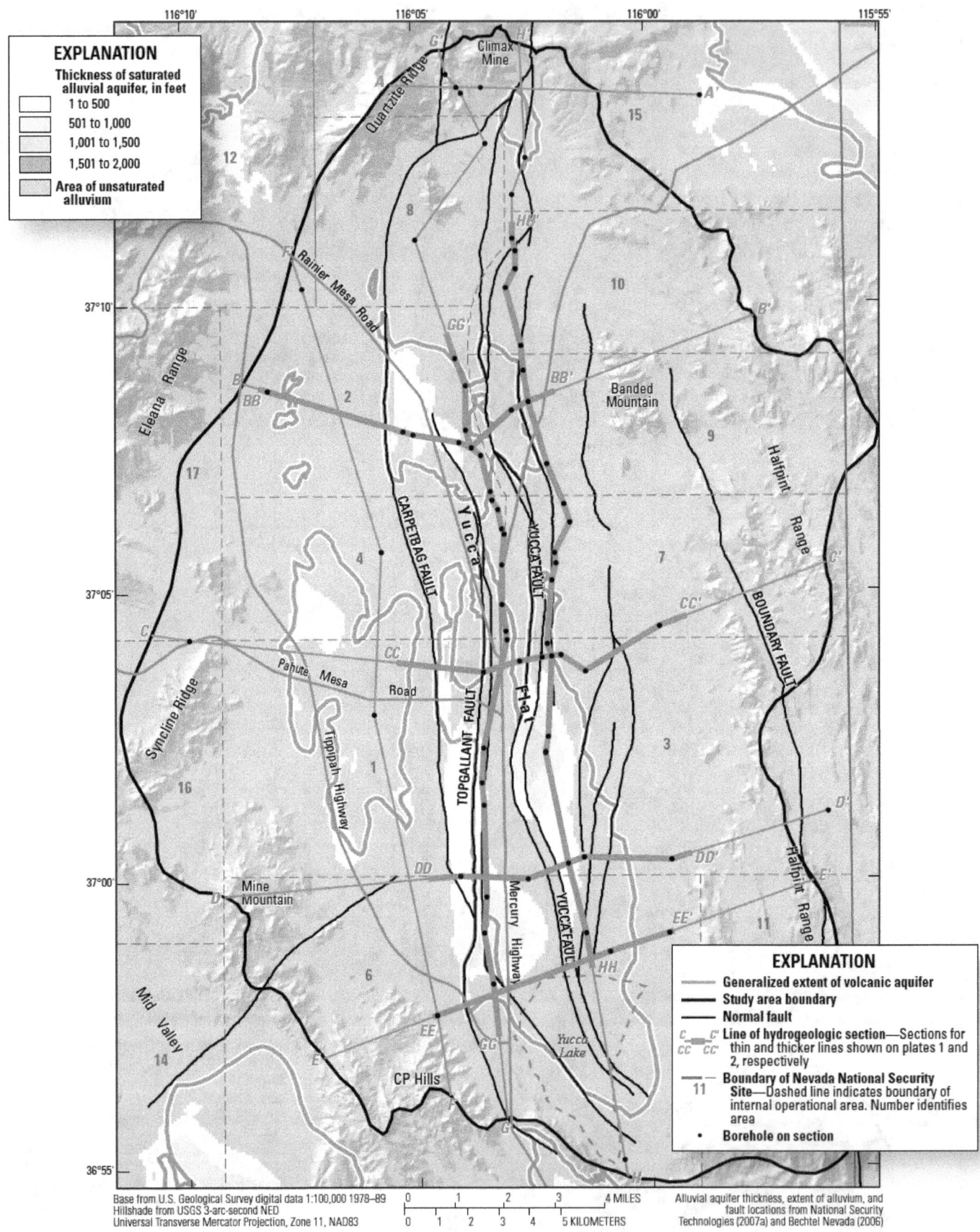

Figure 8.   Thickness of the saturated alluvial aquifer, and extents of the volcanic aquifer and unsaturated alluvium in Yucca Flat, Nevada.

aquifer, but they are relatively restricted areally (Prothro and Drellack, 1997). Multiple volcanic HSUs are mapped in the subsurface of the study area on the basis of geologic formation, vitric/devitrified boundaries, and presence of fractured welded tuff (fig. 4; Bechtel Nevada, 2006). Several of these permeable HSUs stratigraphically overlie each other and form a nearly continuous section that makes up the volcanic aquifer of this report (fig. 7; pls. 1, 2). The continuity of this aquifer is interrupted locally in southern Yucca Flat by a relatively thin volcanic confining unit.

The volcanic aquifer is relatively thick in three areas of Yucca Flat. One area is in the western half of Yucca Flat (fig. 9; sections C–C' and F–F', pl. 1). A second area is in north-central Yucca Flat, where it occurs in the downthrown side of Carpetbag fault (fig. 9; section BB–BB', pl. 2). The aquifer in this area ranges mostly between 0 and 500 ft in thickness, but locally the aquifer is greater than 1,000 ft thick (fig. 9). A third large area of volcanic aquifer is in the southern half of Yucca Flat, where the aquifer is preserved on the downthrown eastern sides of the Yucca and Topgallant faults (fig. 9; section DD–DD', pl. 2). The aquifer in this area is thick and extensive. In large areas the aquifer ranges from 500 to 1,000 ft in thickness, and in local areas near Topgallant fault the aquifer exceeds 1,500 ft in thickness (fig. 9). The saturated part of the volcanic aquifer is sufficiently thick in south-central Yucca Flat that offset on Yucca fault does not disrupt the continuity of the volcanic aquifer (sections DD–DD' and EE–EE', pl. 2). The volcanic aquifer is relatively deep in southern Yucca Flat, but the contact with the underlying volcanic confining unit rises in elevation northward until the confining unit is present at the water table and separates the southern and northern areas of saturated volcanic aquifer (fig. 9; section GG–GG', pl. 2).

In most areas of Yucca Flat, the volcanic aquifer is underlain by less permeable rocks of the volcanic confining unit (fig. 9). Exceptions to this occur in small areas in the western part of Yucca Flat where the volcanic confining unit is absent and the volcanic aquifer directly overlies the carbonate aquifer or the siliceous confining unit.

Volcanic Confining Unit

The lower one-half to two-thirds of the volcanic section in Yucca Flat consists of nonwelded ash-flow tuff, bedded tuff, and tuffaceous sediments that have been variably altered to zeolite and clay minerals as a result of post-volcanic reactions with groundwater (Bechtel Nevada, 2006; Drellack and others, 2010). Zeolitic and argillic alteration result in the occlusion of porosity and a decrease in rock permeability; the alteration can affect the entire unit or parts of multiple stratigraphic units (Winograd and Thordarson, 1975; Laczniak and others, 1996). Although historically considered a single volcanic confining unit (Winograd and Thordarson, 1975; Laczniak and others, 1996), this confining unit section was subdivided on the basis of alteration mineralogy into three separate HSUs and used as such in construction of the Yucca Flat three-dimensional HFM

(fig. 4; Bechtel Nevada, 2006; Drellack and others, 2010). Similar to the older studies, the three HSUs—lower tuff, Oak Spring Butte, and argillic tuff confining units (LTCU, OSBCU, and ATCU on fig. 4)—are considered to have similar bulk hydrologic properties and are not distinguished from each other in this report. These units stratigraphically overlie each other throughout most of the subsurface in Yucca Flat to create a thick, continuous confining unit that may be combined as part of a single SHUT, the volcanic confining unit (fig. 7, pls.1, 2). One additional HSU, the upper tuff confining unit (UTCU on fig. 4), occurs higher in the volcanic-rock sequence as a thin HSU in southern Yucca Flat that locally bisects the volcanic aquifer (sections DD–DD' and EE–EE', pl. 2).

The volcanic confining unit is thickest in the southern half of Yucca Flat where it is commonly greater than 1,500 ft thick on the downthrown eastern sides of Yucca and Topgallant faults (fig. 10; sections DD–DD' and EE–EE', pl. 2). Regional thickness trends show the volcanic confining unit is thinnest adjacent to and between individual strands of Yucca and Topgallant faults where parts of the unit are thinned or omitted by faulting (fig. 10; section CC–CC', pl. 2). In eastern and central Yucca Flat, the volcanic confining unit dips to the west towards these major faults (pls. 1, 2). As a result, the base of the volcanic section increases in elevation eastward such that the saturated volcanic confining unit thins eastward to a point of zero thickness where carbonate rocks are exposed at the water table (figs. 7 and 10; pls. 1, 2).

The volcanic confining unit is an important hydrogeologic unit over much of Yucca Flat because it separates the volcanic aquifer from the underlying carbonate aquifer (pls. 1, 2; Bechtel Nevada, 2006; Drellack and others, 2010). Throughout most of Yucca Flat, the volcanic confining unit has a greater spatial extent than the overlying volcanic aquifer. This enables the volcanic confining unit to provide an effective barrier between the volcanic and carbonate aquifers (fig. 10; pl. 4).

Carbonate Aquifer

The carbonate aquifer SHUT consists of three HSUs: the lower carbonate aquifer, the lower carbonate aquifer thrust plate, and the upper carbonate aquifer (fig. 4). The lower carbonate aquifer is a thick assemblage of early Cambrian and late Middle Devonian, interbedded dolomite and limestone that underlies most of the study area. The lower carbonate aquifer thrust plate HSU consists of thrust-bounded blocks of the same rock units as the lower carbonate aquifer. The upper carbonate aquifer is a locally prominent sequence of Pennsylvanian-aged carbonate rocks underlying Syncline Ridge (fig. 2).

The carbonate aquifer SHUT may be subdivided into local and regional components, referred to in this report as local and regional carbonate aquifers (fig. 3). The classification of a block of carbonate rock as either a regional or a local aquifer is based on the block's lateral and vertical extent and subsurface configuration. Regional carbonate aquifers are

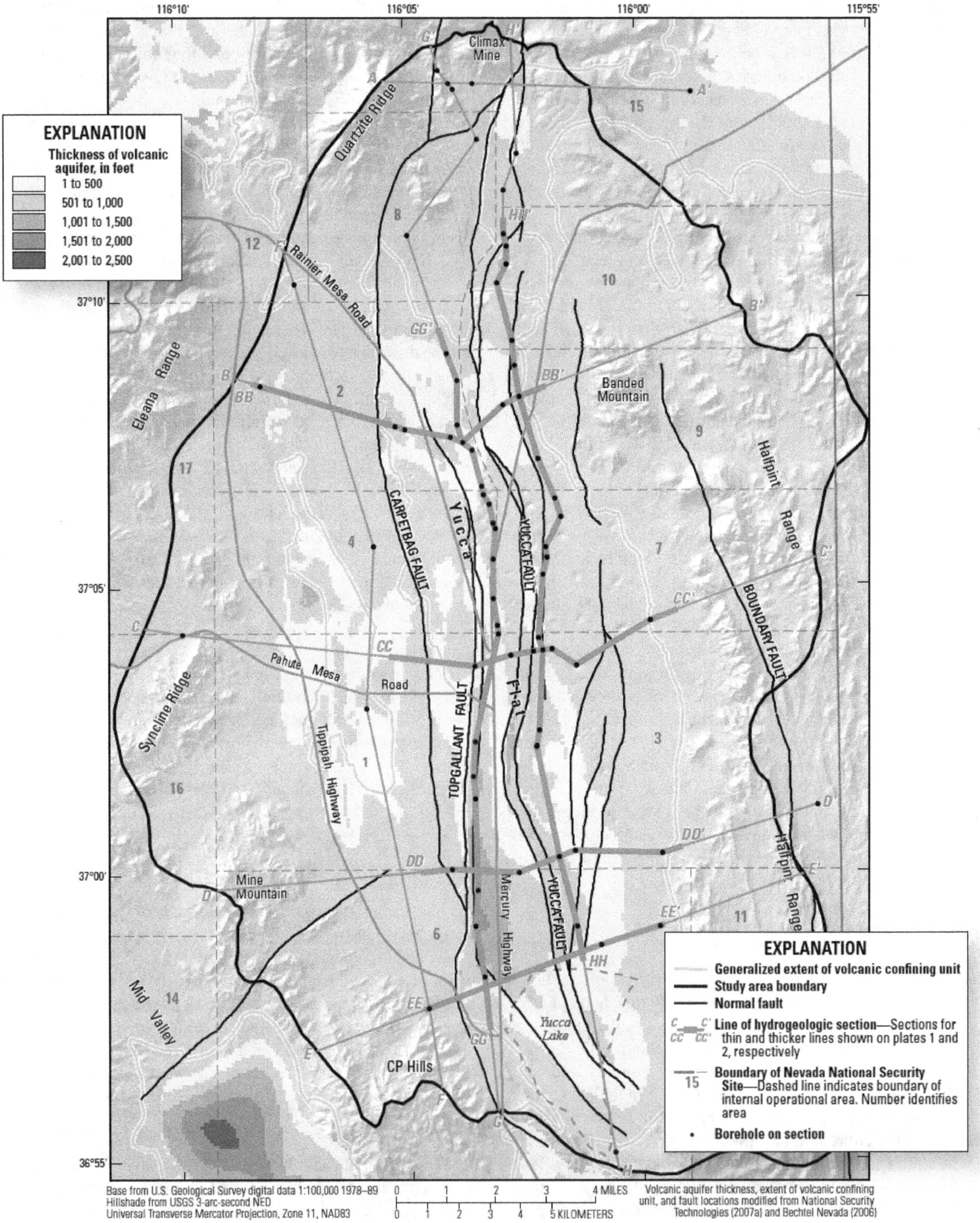

Figure 9.   Thickness of the saturated volcanic aquifer and extent of the underlying volcanic confining unit in Yucca Flat, Nevada.

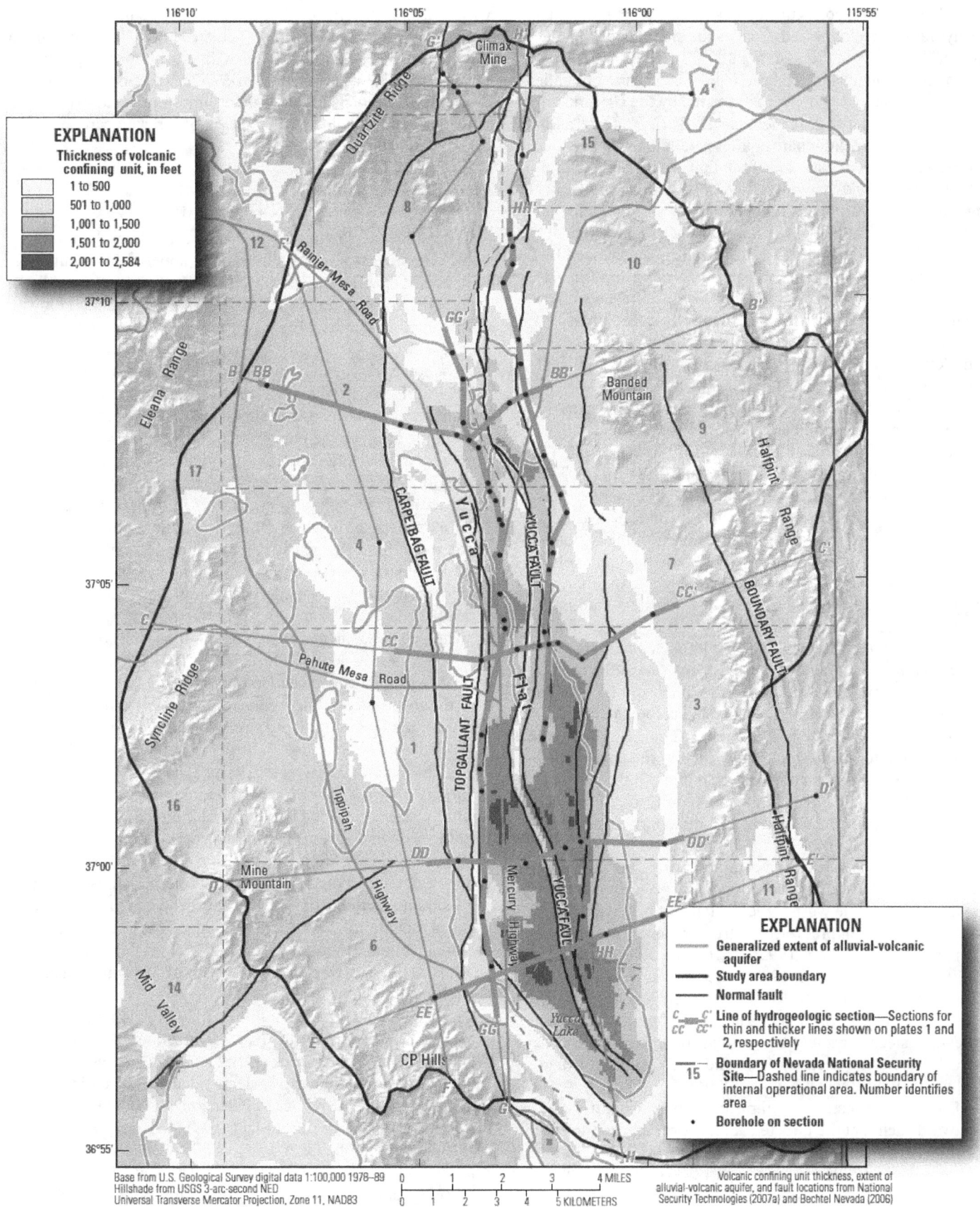

Figure 10.   Thickness of the saturated volcanic confining unit and extent of overlying alluvial–volcanic aquifer in Yucca Flat, Nevada.

laterally extensive, contiguous blocks of carbonate rock that are hydraulically connected, such as the saturated carbonate aquifer that exists in the eastern half of Yucca Flat (fig. 7). Regional carbonate aquifers also may exist as thrust-bounded slices, such as the carbonate rock carried by the CP thrust in western Yucca Flat (fig. 11; sections *B–B', C–C', D–D'* and *F–F',* pl. 1). Local carbonate aquifers are less extensive, disconnected blocks of carbonate rock that are isolated, hydraulically restricted, and generally drain only to adjacent confining units. Local carbonate aquifers may be stratigraphically isolated, such as in the western part of Yucca Flat where the upper carbonate-rock aquifer HSU is mostly surrounded by the stratigraphically underlying siliceous confining unit (fig. 11; western part of section *C–C',* pl. 1). Local aquifers also may be structurally isolated, such as structurally dismembered blocks associated with the CP thrust that are thrust over, and completely surrounded by, underlying siliceous rock (fig. 11; western part of section *B–B'* near well *UE-2ce,* pl. 1).

The regional carbonate aquifer is present at the water table beneath the eastern part of Yucca Flat, but the aquifer is confined in central Yucca Flat, where it is buried beneath the volcanic section (fig. 7; sections *C–C'* and *D–D',* pl. 1). The aquifer is continuous under the main areas of underground testing in central Yucca Flat (Cole, 1997; Cole and Cashman, 1999). The aggregate stratigraphic thickness of the regional carbonate aquifer is as much as 15,000 ft, although the effective saturated thickness is spatially variable due to thrust repetition or tilting, and to subsequent extensional faulting. The northeastern limit of the saturated part of this aquifer is defined by the depositional contact with the underlying siliceous confining unit beneath northeastern Yucca Flat (fig. 7). The western limit of the saturated part of the aquifer is in part defined by the western extent of thrusted carbonate in the CP thrust plate (fig. 7; sections *B–B'* and *C–C',* pl. 1). The regional carbonate aquifer is inferred to underlie the siliceous confining unit at depth beneath the western margin of the study area (fig. 11; western parts of sections *B–B'* and *C–C',* pl. 1) and occur beneath the CP thrust in the southwestern part of the study area (sections *D–D'* and *F–F',* pl. 1).

The extent and subsurface configuration of the carbonate aquifer in the study area is constrained by pre-Cenozoic thrust faults and folds. The west-vergent CP thrust is the most hydrologically significant pre-Tertiary structure in the study area (Caskey and Schweikert, 1992; Trexler and others, 1996; Cole, 1997; Cole and Cashman, 1999). The sinuous trace of the CP thrust indicates that the thrust is a relatively low-angle fault beneath the western portion of Yucca Flat, but the fault plane appears to steepen rapidly eastward as it approaches central Yucca Flat (fig. 11; sections *C–C'* and *D–D',* pl. 1; Cole and Cashman, 1999; Bechtel Nevada, 2006).

The CP thrust carries the thrusted carbonate aquifer westward such that the thrusted aquifer overlaps non-thrusted regional carbonate aquifer (fig. 11; section *D–D',* pl. 1) or the siliceous confining unit (northern part of section *F–F',* pl. 1). In places the thrusted carbonate rocks are separated from underlying non-thrusted carbonate aquifer by siliciclastic rock carried as the lowest part of the thrusted sequence (fig. 11; section *E–E',* pl. 1; Caskey and Schweikert, 1992; Cole and

Cashman, 1999). The local occurrence of a confining unit at the base of regional thrust sheets creates internal barriers in the regional carbonate aquifer that can divert or separate groundwater flow. Elsewhere, such as at *UE-2ce* (section *B–B',* pl. 1), dismembered blocks of dolomite or limestone lie above the siliceous confining unit and the block of thrusted carbonate is completely isolated.

Siliceous Confining Unit

The siliceous confining unit SHUT consists of three principal HSUs: lower clastic confining unit, upper clastic confining unit, and Mesozoic granite confining unit (fig. 4). Two additional HSUs, consisting of thrust sheets of the upper and lower clastic confining units, also are part of the siliceous confining unit SHUT (fig. 4).

The lower clastic confining unit HSU consists of up to 10,000 ft of Late Proterozoic to Lower Cambrian quartzite, micaceous quartzite, and siltstone. Locally, thin limestone units occur in the upper part of the section. This HSU makes up a basal siliceous confining unit that functions as the regional hydrologic basement throughout the study area because of its generally low water-transmitting characteristics. Where exposed at the water table by anticlinal uplift in the northern Halfpint Range (fig. 7), the siliceous confining unit impedes the movement of groundwater and results in steep horizontal hydraulic gradients (Winograd and Thordarson, 1975, pl. 1).

The upper clastic confining unit HSU consists of as much as 6,500 ft of low-permeability Late Devonian and Mississippian-aged siliceous siltstone, sandstone, conglomerate, and minor limestone. The saturated part of this unit occurs at the water table as a relatively narrow band along the western edge of the study area (fig. 7). Structurally, this unit exists within the synclinal downwarp at Syncline Ridge (fig. 7, section *C–C',* pl. 1). These rocks underlie a local carbonate aquifer in the core of the fold near borehole UE-16d WW (fig. 11; section *C–C',* pl. 1). Because of the large thickness of upper clastic confining unit rocks in the downwarp, the top of the regional carbonate aquifer can occur at depths exceeding 9,000 ft (section *C–C',* pl. 1). This deep trough of clastic rocks is thought to separate groundwater flow in the shallow regional carbonate aquifer in Yucca Flat from flow in the shallow regional carbonate aquifer west of Syncline Ridge (Fenelon and others, 2010).

The Mesozoic granite confining unit HSU consists of granodiorite and porphyritic quartz monzonite that comprise the Cretaceous Climax stock (U.S. Geological Survey, 1983). The Climax stock is a narrow, steep-sided, nearly cylindrical intrusive body that invades the Paleozoic carbonate section at the northern end of Yucca Flat (fig. 7; Houser and Poole, 1960; Phelps and others, 2004; Bechtel Nevada, 2006). The granitic rocks have a low primary porosity and permeability, but small quantities of water may pass through these rocks where fractures or weathered zones exist. However, the fractures are poorly connected, and these rocks generally impede groundwater flow (Winograd and Thordarson, 1975).Where

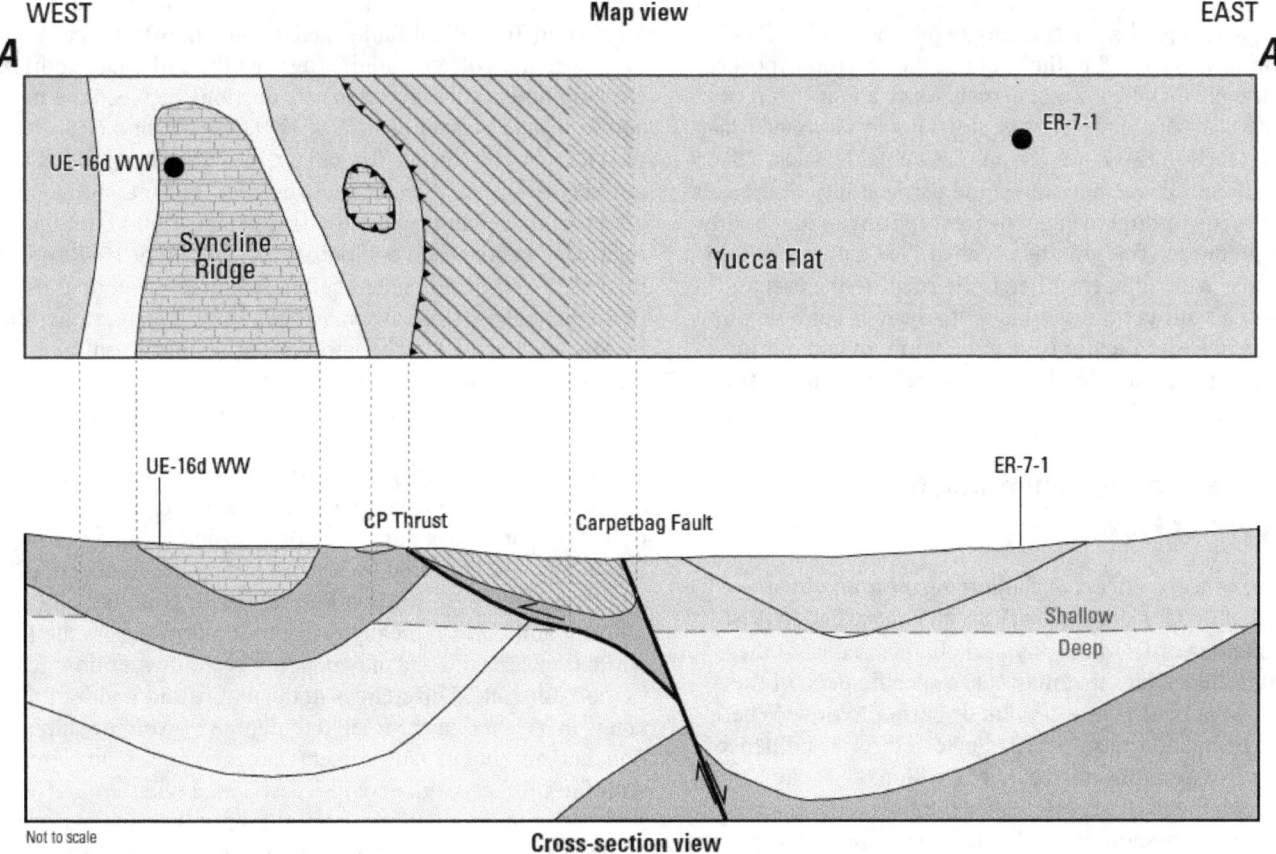

WEST   Map view   EAST

A   **A′**

Section and map portray generalized pre-Cenozoic geologic setting. Cenozoic geologic units and structures, with the exception of the Carpetbag fault, are not shown. Modified from Fenelon and others (2010).

EXPLANATION

Regional carbonate aquifer

Shallow part—Area where flow is most active and where top of saturation is within about 6,000 ft of land surface

Thrusted—Area where shallow carbonate aquifer is thrust over another shallow carbonate aquifer and the two aquifers are separated by siliciclastic rock. Pattern shown on map view only

Thrusted—Area where shallow carbonate aquifer is thrust over a deep carbonate aquifer and the two aquifers are separated by siliciclastic rock. Pattern shown on map view only

Thrusted—Area of undifferentiated thrusted carbonate aquifer. Pattern shown on cross-section view only

Deep part—Area where flow is assumed less active and where top of saturation is greater than about 6,000 feet below land surface

Shallow / Deep   Approximate boundary between shallow and deep parts of the regional carbonate aquifer

Local carbonate aquifer

Approximate saturated extent of localized blocks of moderately permeable carbonate rock

Siliceous confining unit

Approximate saturated extent of siliceous rocks of low permeability. Includes upper and lower clastic confining units

Thrust fault—Map trace; teeth on upthrown block

Thrust fault—On cross section; arrow on upthrown block shows direction of motion

Normal fault—On cross section; arrows show sense of offset

Form line—Shows general attitude of bedding in cross-section view

Construction line—Shows corresponding features in map and cross-section view

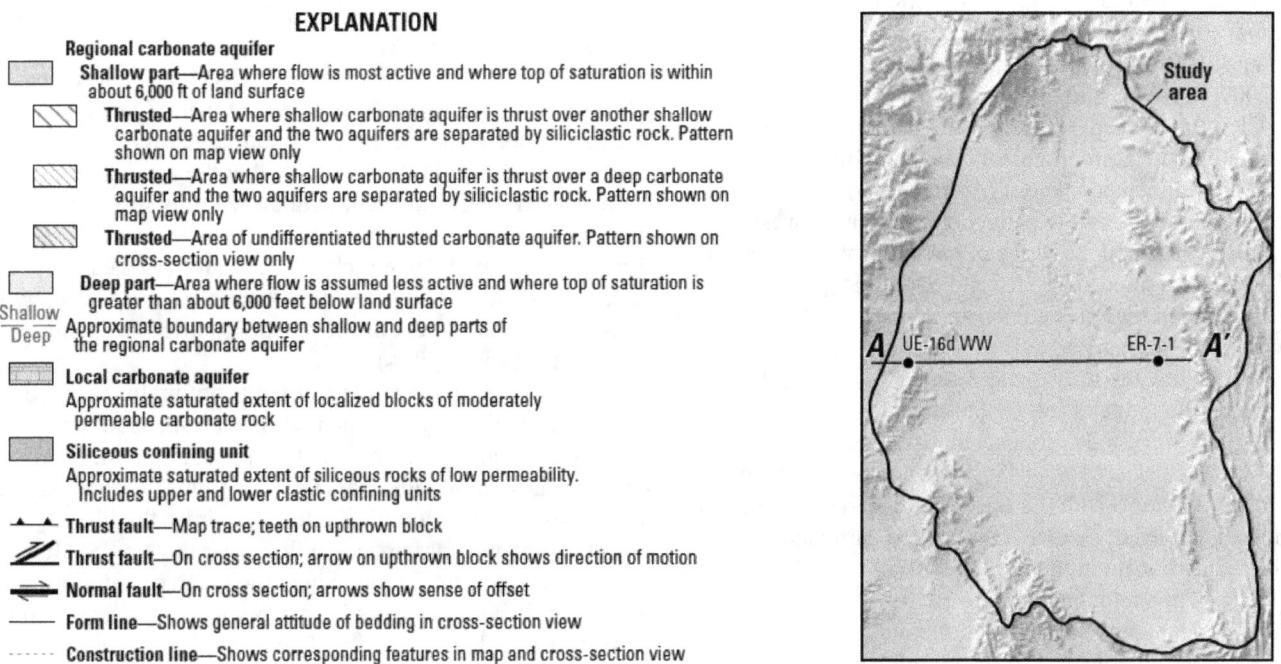

Figure 11.   Generalized hydrogeologic map and cross section for pre-Cenozoic rocks in Yucca Flat.

the intrusive rocks invade the subsurface, little or no carbonate aquifer is anticipated to exist at any depth. Borehole ER-8-1, located in NNSS Area 8 to the south of the outcrop exposures of the Climax stock (fig. 2), penetrated only about 700 ft of unsaturated carbonate rock before finishing in saturated granitic rock (section G–G', pl. 1; app. 3; Bechtel Nevada, 2004). The granitic intrusions may lower the permeability of the adjacent host rocks through contact metamorphism and by hydrothermal alteration. The granitic rocks of the Climax stock, in combination with adjacent siliciclastic rocks to the east and west, form a semi-continuous band of siliceous confining unit that serves to hydraulically isolate the northern end of Yucca Flat from aquifers outside of the study area to the north (fig. 7; Bechtel Nevada, 2006).

Geologic Structures Important to Groundwater Flow

The hydrologic effects of faulting result from either fault-caused juxtaposition of rocks with contrasting hydrologic properties or from the physical characteristics of the fault zones themselves that may cause specific parts of the fault zone to act either as a conduit or barrier to flow. Where faults are open and transmissive, they can directly influence flow rates and groundwater velocities within the faulted zone; alternatively, the fault may influence the flow field in a region surrounding or enveloping the faulted zone (Black and others, 1987). The following factors bear on the influence a fault zone has on the local flow field: (a) the dimensions and character of the fault core and damage zones; (b) the type of rocks present on both sides of the fault; and (c) orientation of the fault with respect to the modern-day stress field (Hansen and others, 1963; Faunt, 1997; Ferrill and others, 1999; Sweetkind and Drake, 2007a; Prothro and others, 2009).

The CP thrust is responsible for the juxtaposition of large blocks of carbonate aquifer and siliceous confining unit. The resulting three-dimensional configuration of aquifers clearly influences groundwater flow paths in local areas in western Yucca Flat, despite a high degree of uncertainty in the definition of the potentiometric surface and hydraulic gradients in this fault-bounded region (Fenelon and others, 2010). The siliceous confining unit carried at the base of the CP thrust sheet, or occurring below the thrust, was interpreted to partially or fully separate the thrusted regional carbonate aquifer from the underlying, non-thrusted carbonate aquifer (sections B–B' and E–E', pl. 1). The eastern, downward-steepening part of the CP thrust, in combination with the presence of Carpetbag fault, was inferred to partially isolate the thrusted carbonate from non-thrusted carbonate in the eastern half of Yucca Flat (section C–C', pl. 1; Fenelon and others, 2010).

Yucca, Topgallant, Carpetbag, and related smaller faults have juxtaposed the volcanic confining unit and volcanic aquifer by hundreds of feet (pls. 1, 2). These faults locally compartmentalize the volcanic aquifer into north-south blocks in Yucca Flat. An example is shown on section CC–CC' (pl. 2), where the volcanic aquifer is bounded on the east and west by

fault contacts with the volcanic confining unit. Additionally, Yucca and Topgallant faults have sufficient offset locally to juxtapose the volcanic aquifer against the carbonate aquifer, creating potentially important connections between the two aquifer systems (sections CC–CC', DD–DD', and GG–GG', pl. 2). Offset by Yucca, Topgallant, and Carpetbag faults is unlikely to isolate the carbonate aquifer. The magnitude of offset on these faults generally is less than 1,000 ft and is relatively minor when compared to the 6,000 to 10,000-foot thickness of the carbonate aquifer. As such, offset on these faults typically place carbonate against carbonate, rather than juxtapose the entire carbonate aquifer against a confining unit (sections C–C' and E–E', pl. 1).

Field observations of faults in the vicinity of Yucca Flat (Hansen and others, 1963; Sweetkind and Drake, 2007a; Prothro and others, 2009) suggest that, like faults described elsewhere, they are zoned into a fault core composed of clay-rich gouge or matrix-supported breccia and a damage zone of brecciated and fractured rock surrounding the fault core (Caine and others, 1996; Caine and Forster, 1999; Kim and others, 2004). Fault cores typically restrict fluid flow across the fault, while the damage zone may conduct groundwater flow parallel to the fault zone. Differences in the nature and width of these zones in Yucca Flat are related to degree of welding, alteration, and amount of fault offset. Damage zones tend to scale with fault offset; damage zones associated with large-offset faults (greater than 300 ft) are 100 ft or more wide, whereas damage zones associated with smaller offset faults are generally only a few feet wide (Hansen and others, 1963; Sweetkind and Drake, 2007a; Prothro and others, 2009).

Faults also may influence the flow of groundwater through the regional carbonate aquifer. Field observations of fault zones in carbonate rocks in the vicinity of Yucca Flat document 30- to 60-foot wide damage zones adjacent to faults (Prothro and others, 2009). A large number of subsidiary normal faults occur east of Yucca fault in eastern Yucca Flat (section D–D', pl. 1). It is possible that the aggregate width of fractured rock associated with these faults sufficiently enhances the permeability of the carbonate aquifer to create a broad zone of carbonate rock with elevated permeability. The existence of such a zone would help explain the extremely low north-to-south gradient of the potentiometric surface in east-central Yucca Flat (pl. 4).

Seismically active faults and faults optimally oriented for failure with respect to the present-day stress field may be of special interest from a hydrologic standpoint. Faunt (1997) analyzed in-situ stress measurements, earthquake focal mechanisms, and geologic evidence, to infer the likelihood of faults as conduits or barriers to flow near the NNSS. Given the present-day stress field, where the mean orientation of the minimum horizontal stress is approximately northwest-southeast (Stock and others, 1985), Faunt (1997) suggested that faults in relative tension (north- to northeast-striking) would be conduits for flow, and those in relative compression (northwest-striking) would be barriers to flow. As such, all of the north-striking normal faults in Yucca Flat would be expected to be in relative tension and behave as conduits for flow.

Predevelopment Flow

The alluvial–volcanic and carbonate aquifer systems make up the groundwater flow systems in Yucca Flat. The alluvial–volcanic aquifer system consists of the alluvial–volcanic aquifer and the volcanic confining unit (fig. 3). Underlying and surrounding this aquifer system is the carbonate aquifer system, which consists of the carbonate aquifer and the siliceous confining unit.

Groundwater flow within Yucca Flat is portrayed using contours drawn on the basis of hydraulic heads interpreted to be representative of predevelopment conditions and from the known or inferred distribution of aquifers and confining units, major structures that are likely to function as barriers or conduits, and locations of recharge and discharge. Predevelopment conditions are intended to represent the state of the groundwater system prior to any changes introduced by human activity. The contours are shown for the alluvial–volcanic aquifer system on plate 3 and for the carbonate aquifer system on plate 4. These contours are used to infer directions of groundwater flow in the aquifer systems.

Hydraulic heads specifically used in the contouring process, as well as additional heads used to constrain contours, are posted on the plates and detailed in appendix 2. Hydraulic heads representative of the alluvial–volcanic aquifer and volcanic confining unit are posted on plate 3 and heads representative of the carbonate aquifer and siliceous confining unit are posted on plate 4. The specific subsurface hydrologic unit type (SHUT) to which each well is open is identified by the well symbol type and color on the plates. Hydraulic heads denoted next to a triangle symbol and shown in a grey font are composite heads measured in wells open to multiple SHUTs. Composite heads consistent with heads in the alluvial–volcanic aquifer system are shown on plate 3, whereas composite heads that appear to be influenced more by the carbonate aquifer system are shown on plate 4. The font size of the posted hydraulic heads (pls. 3 and 4) indicates the level of certainty in which the hydraulic head represents predevelopment conditions—the larger the font size, the higher the level of certainty. Hydraulic heads shown in italics are considered anomalous and were not used in developing contours.

The hydraulic gradient, which influences the direction and rate of groundwater flow and contaminant movement within an aquifer, can be approximated from spatial differences between contours of the potentiometric surface. The general flow direction, as defined by these contours, is shown on plates 2, 3, and 4 by arrows within the mapped extent of each major aquifer. Small and intermediate arrows along aquifer boundaries indicate likely directions of leakage to or from adjacent confining units or lateral flow to or from one aquifer to another, respectively.

Contours presented in this report for predevelopment potentiometric surfaces (pls. 3, 4) are assumed also to approximate modern-day, post-development, conditions. This is because most of the potentiometric response in the major aquifers to pumping or past nuclear testing is interim and localized. Persistent, long-lasting (greater than 50 years) changes in hydraulic head caused by nuclear testing occur only in confining units where groundwater fluxes are negligible. These transient changes have only a minimal influence on the general flow directions in the aquifers shown on the plates. Pumping has caused widespread small (less than several feet) changes in hydraulic head or larger localized drawdowns. In either case, this influence on large-scale flow in Yucca Flat is minimal.

Alluvial–Volcanic Aquifer System

The alluvial–volcanic aquifer system consists of the alluvial aquifer, volcanic aquifer, and volcanic confining unit (figs. 8–10). The alluvial aquifer is assumed to be hydraulically connected vertically and laterally to the underlying volcanic aquifer in Yucca Flat (sections *BB–BB'*, *DD–DD'*, *GG–GG'*, and *HH–HH'*, pl. 2). A direct hydraulic connection is supported not only by the physical connection between the two aquifers, but by hydraulic heads that show little variation between the aquifers (pl. 3). Because of the hydraulic connection, these aquifers are grouped together and referred to as the alluvial–volcanic aquifer.

The alluvial–volcanic aquifer occurs at the water table and is unconfined throughout its extent. Locally, the aquifer may be confined or semi-confined at depth where a thin layer of the volcanic confining unit occurs within the aquifer, such as near UE-6e (section *HH–HH'*, pl. 2). The sides and bottom of the alluvial–volcanic aquifer are almost completely bounded by the volcanic confining unit (fig. 10; pl. 2). Only in local areas does the aquifer directly overlie or abut the regional carbonate aquifer. These areas occur primarily between Carpetbag and Topgallant faults in Area 1 of the NNSS (fig. 10; southern end of section *DD–DD'*, pl. 2); just east of Carpetbag fault in south-central Area 2 (fig. 10); west of Carpetbag fault in central Area 1 (fig. 10; section *C–C'*, pl. 1); and in central Yucca Flat where Yucca fault juxtaposes alluvial–volcanic aquifer with carbonate aquifer (section *CC–CC'*, pl. 2).

The alluvial–volcanic aquifer system is considered a local flow system (Fenelon and others, 2010). Water within it cannot flow directly to a downgradient point of discharge at land surface. Rather, groundwater within this system is isolated by the volcanic confining unit that underlies the aquifer nearly throughout its entire extent. All discharge from this local flow system is controlled by hydraulic connections and local hydraulic gradients between the alluvial–volcanic and carbonate aquifers. Vertical hydraulic gradients between the alluvial–volcanic aquifer and the regional carbonate aquifer generally indicate downward flow potential (pl. 4). Any groundwater moving downward into the carbonate aquifer through the confining unit is expected to be limited by the impeding nature of the confining unit. The total downward flux from the alluvial–volcanic aquifer system into the carbonate aquifer system was estimated to be 25–65 acre-ft/yr (Winograd and Thordarson, 1975). Discharge from the alluvial–volcanic aquifer system occurs over a broad area as vertical diffuse leakage through the volcanic confining unit or as focused drainage through faults or where the alluvial–volcanic and carbonate aquifers are in direct contact. Local areas of contact occur along the boundary of the alluvial–volcanic aquifer where no volcanic confining unit intervenes or in isolated locations where a fault juxtaposes aquifer against aquifer.

Groundwater flow in the aquifers within the alluvial–volcanic aquifer system is discussed by region. One aquifer region, located west of Carpetbag fault in NNSS Areas 1 and 4 (pl. 3), herein is referred to as the *western alluvial–volcanic aquifer*. Two additional regions occur in Yucca Flat proper, within the main part of the aquifer system east of Carpetbag fault. The aquifer within the northern part of the system, which includes Areas 2 and 9 of the NNSS, is referred to as the *northern alluvial–volcanic aquifer*. The aquifer within the southern part of the system, which includes Areas 1, 3, 4, 6, and 7, is referred to as the *southern alluvial–volcanic aquifer*. A final region in the far northern part of the study area in the southwestern part of Area 15 of the NNSS contains volcanic confining unit and lesser amounts of alluvial–volcanic aquifer. This region has limited hydraulic information and is represented by a single 2,520-ft contour. The flow interpretation for this region is limited to suggesting southwesterly flow toward the regional carbonate aquifer. In the northern part of this region, the volcanic confining unit is thought to support a perched water table that overlies a regional water table in the regional carbonate aquifer (sections A–A' and H–H', pl. 1). Because this far northern region is not the focus of the study, it is not discussed further.

Western Alluvial–Volcanic Aquifer

The western alluvial–volcanic aquifer covers a large area west of Carpetbag fault (pl. 3). No nuclear devices were detonated in this aquifer, although several tests were conducted just to the north in Areas 2 and 4 of the NNSS (fig. 7). Most of the aquifer is bounded by the siliceous confining unit (western part) and the volcanic confining unit (central part). However, the alluvial–volcanic aquifer directly overlies thrusted regional carbonate aquifer along its northeastern and southern ends and abuts local carbonate aquifer at Syncline Ridge on its northwestern edge. These relations are apparent on section C–C', east of borehole UE-16d WW (pl. 1). From west to east, this section depicts the alluvial–volcanic aquifer (1) laterally connected to a thin section of local carbonate aquifer at a high in the siliceous confining unit; (2) directly underlain by the siliceous confining unit; (3) directly underlain by the volcanic confining unit; and (4) directly underlain by thrusted regional carbonate aquifer.

Water in the western alluvial–volcanic aquifer is conceptualized as flowing generally eastward. Water is derived from lateral flow into the western side of the aquifer, primarily from the local carbonate aquifer but also with seepage from the siliceous confining unit. The source of the water is recharge from precipitation falling on highlands at Syncline Ridge and Mine Mountain (pl. 3). Water flows eastward through the aquifer

and discharges primarily into the thrusted regional carbonate aquifer.

The flow conceptualization is uncertain and is based on one hydraulic head from well *UE-1c* (pl. 3) and general hydrologic concepts. These concepts include: (1) a recharge source to the west; (2) the general assumption of west-to-east flow in the area (Fenelon and others, 2010), based on nearby hydraulic heads suggesting a large head drop between Syncline Ridge and central Yucca Flat (pls. 3 and 4); and (3) direct connections between the alluvial–volcanic aquifer and the thrusted regional carbonate aquifer to the east that provide a drain for water. Well *UE-1c* is open to 460 ft of volcanic aquifer and 108 ft of carbonate aquifer and has a hydraulic-head estimate of 2,909 ft (app. 2). The relatively high head value is assigned to the alluvial–volcanic aquifer because of the large thickness of volcanic aquifer at the open interval, characteristics of the water chemistry in well *UE-1c* that suggest the water has a volcanic-rock chemical signature (Farnham and others, 2006), and because the alluvial–volcanic aquifer is more likely to have an elevated head than the underlying carbonate aquifer. This lone head value is used to construct a single north-south 2,900-ft contour. The contour implies that flow generally is eastward through the extent of the aquifer (pl. 3) and that the potentiometric surface in the aquifer is elevated above heads in the underlying regional carbonate aquifer. The conceptualization is based on two assumptions. First, the head in well *UE-1c* is assumed to represent the volcanic aquifer rather than the underlying carbonate aquifer. Second, the thin volcanic confining unit at the base of the volcanic aquifer is assumed to be an effective barrier that can support a hydraulic head that is elevated about 400 ft above the underlying regional carbonate aquifer. If the water table is elevated in this thin alluvial–volcanic aquifer system, then perched water may occur over much of the area underlain by regional carbonate aquifer (section C–C' and F–F', pl. 1).

Southern Alluvial–Volcanic Aquifer

The southern alluvial–volcanic aquifer is located in the southern part of Yucca Flat proper. This aquifer is bounded by Carpetbag fault to the west and encompasses Yucca Lake to the south and the southern part of NNSS Areas 4 and 7 to the north (pl. 3). The alluvial–volcanic aquifer is unconfined and is bounded below and laterally on the northern, eastern, and southern ends by the volcanic confining unit (fig. 10; pl. 3). The far western side of the aquifer directly overlies or abuts the regional carbonate aquifer (sections CC–CC', DD–DD', and EE–EE', pl. 2). A large number of underground nuclear tests were conducted in and near the southern alluvial–volcanic aquifer (fig. 7; sections CC–CC', DD–DD' and southern two-thirds of GG–GG' and HH–HH', pl. 2). Most of these tests were done in the northern two-thirds of the aquifer and east of Yucca fault. The aquifer is dissected by normal faults including Yucca fault, which bisects the aquifer into western and eastern parts (pl. 3). Yucca fault is hydrologically important because the large offset on the fault locally juxtaposes alluvial–volcanic aquifer against regional carbonate aquifer in the vicinity of borehole TW-7 (pl. 3; section *CC–CC'*, pl. 2). This important but limited area of hydraulic connection with the regional carbonate aquifer, in combination with isolation of most of the aquifer by the volcanic confining unit, provide the principal controls on flow in the southern alluvial–volcanic aquifer.

Hydraulic heads contoured in the southern alluvial–volcanic aquifer range in value from greater than or equal to 2,533 ft at well *UE-3e 4-3 (1661 ft)* to 2,399 ft at well *U-3jn 1* (pl. 3; app. 2). Contours developed from these heads range from 2,530 to 2,400 ft. Hydraulic heads are elevated 20 to 100 ft above the regional carbonate aquifer throughout most of the southern alluvial–volcanic aquifer (pls. 3 and 4).

Water enters the aquifer laterally through leakage from the bounding volcanic confining unit and from direct flow into the aquifer on its southwestern boundary from the regional carbonate aquifer (pl. 3). A hydraulic connection is likely along the southwestern boundary where hydraulic heads in the alluvial–volcanic and carbonate aquifers are nearly identical. The head similarity is shown in a comparison of wells *WW-3 (1800 ft)*, open to the alluvial–volcanic aquifer, and *UE-1h*, open to the regional carbonate aquifer; these wells are located less than 1 mi apart. The head in well *WW-3 (1800 ft)* is 2,437 ft (pl. 3; app. 2), whereas the head in well *UE-1h* is 2,440 ft (pl. 4; app. 2). Direct recharge to the southern alluvial–volcanic aquifer and bounding volcanic confining unit from infiltration of modern-day precipitation is negligible (0 to 0.004 in/yr; Hevesi and others, 2003, fig. 43). Elevated heads in the alluvial–volcanic aquifer, despite no significant modern-day recharge, are believed to reflect slow equilibration from higher water tables that occurred during a period of wetter climatic conditions. This concept is discussed in more detail in the "Volcanic Confining Unit" section of the report.

Groundwater in the aquifer is portrayed as flowing in a U-shaped, counter-clockwise direction. Groundwater flows generally in a southerly direction on the western side of Yucca fault, crosses the fault in the area of borehole U-3jg, and then flows in a northerly direction on the eastern side of the fault (pl. 3). The primary driver for this reversal in flow direction between the western and eastern parts of the aquifer is Yucca fault. The fault creates a barrier to flow in the northern part of the southern alluvial–volcanic aquifer, and at the same time provides a conduit for flow out of the aquifer. The mechanism for these two opposing flow dynamics is thought to be controlled primarily by simple juxtaposition of rock units rather than by the hydraulic properties of the fault itself. Yucca fault is conceptualized to create a barrier to eastward flow on the west side of the fault by either physically separating the western and eastern parts of the aquifer with a wedge of the volcanic confining unit (pl. 3; section *CC–CC'*, pl. 2) or by thinning the western part of the aquifer to the point that the hydraulic connection from west to east is poor (fig. 9). Hydraulic heads west of the fault are higher than heads on the east side. For example, the head in well *UE-1k* is 2,447 ft, whereas directly to the east, the head in well *ER-3-2-1 (deep)*

is 2,404 ft (pl. 3; app. 2). Further south, head measurements are similar across the fault, suggesting that the fault provides little impedance to eastward flow. On the east side of the fault, heads decline northward from 2,428 ft at well *TW-B* to 2,399 ft at well *U-3jn 1* (pl. 3; app. 2).

The lowest heads in the aquifer occur in the vicinity of wells *U-3jn 1* and *TW-7*, coinciding with an area along Yucca fault where the alluvial–volcanic aquifer is believed to be hydraulically connected to the underlying regional carbonate aquifer (pl. 3; section *CC–CC'*, pl. 2; Winograd and Thordarson, 1975; Bechtel Nevada, 2006; Fenelon and others, 2010). This hydraulic connection provides the primary drain in the southern alluvial–volcanic aquifer. Hydraulic heads in the vicinity of the drain (wells *TW-7* and *U-3jn 1*) are nearly identical to heads directly below in the regional carbonate aquifer (pls. 3, 4), although presumably heads are slightly lower in the carbonate aquifer.

Outflow from the southern alluvial–volcanic aquifer, south of borehole UE-4ae and in the area of Carpetbag fault, is shown on plate 3. A small amount of outflow is assumed in this area because the alluvial–volcanic aquifer has been mapped to directly overlie the carbonate aquifer and limited heads in the two aquifers indicate a hydraulic gradient toward the carbonate aquifer (pls. 3, 4). The interpretation of outflows across this boundary is uncertain.

Northern Alluvial–Volcanic Aquifer

The northern alluvial–volcanic aquifer covers the northern part of Yucca Flat proper, primarily in NNSS Areas 2 and 9. This relatively thin water-table aquifer is bounded below and laterally on its northern, eastern, and southern ends by the volcanic confining unit (pl. 3, sections *BB–BB'* and *GG–GG'*, pl. 2). On its western side, the aquifer abuts thrusted regional carbonate aquifer (section *BB–BB'*, pl. 2). In a small area just east of Carpetbag fault, the alluvial–volcanic aquifer is mapped as lying directly on top of the carbonate aquifer. Numerous underground nuclear tests were conducted in the vicinity of the northern alluvial–volcanic aquifer (fig. 7; sections *BB–BB'* and northern parts of *GG–GG'* and *HH–HH'*, pl. 2).

Hydraulic heads contoured in the northern alluvial–volcanic aquifer range from 2,505 ft at well *UE-2fb* to 2,424 ft at well *UE-2aa (2207 ft)* (pl. 3; app. 2). Heads are contoured with a 20-ft contour interval because of a relatively large horizontal hydraulic gradient and a high variability in posted heads on plate 3. Despite the variability, groundwater in the aquifer appears to be flowing primarily to the north with a westward component on the western part of the aquifer. Recharge to the aquifer is derived from lateral leakage from the surrounding volcanic confining unit. Direct recharge to the northern alluvial–volcanic aquifer and bounding volcanic confining unit from infiltration of modern-day precipitation is negligible (0 to 0.004 in/yr; Hevesi and others, 2003, fig. 43). The decline in heads towards the north-central part of the aquifer suggests the likelihood of a connection to the underlying regional carbonate aquifer in this area. The thickness of the saturated volcanic confining unit in the general area of the hydraulic low is estimated to be less than 500 ft (fig. 10; northern end of section *GG–GG'*, pl. 2; Bechtel Nevada, 2006). For example, at borehole UE-2aa, only about 140 ft of the volcanic confining unit separates the alluvial–volcanic aquifer from the underlying carbonate aquifer (pl. 3; app. 3). The hydraulic head in the shallow well *UE-2aa (2207 ft)*, open to the volcanic aquifer, is 2,424 ft. In the deeper well *UE-2aa (2317 ft)*, open to a composite of volcanic aquifer, volcanic confining unit, and carbonate aquifer, the hydraulic head is 2,409 ft (pl. 4; app. 3). The head in the deeper well appears to represent the head in the regional carbonate aquifer in the area of the hydraulic low. Yucca fault could be a pathway for vertical drainage from the alluvial–volcanic aquifer to the carbonate aquifer. Evidence for this is a potential connection between the two aquifers on the eastward fault strand near borehole U-9bx (section *BB–BB'*, pl. 2) and a low hydraulic head near Yucca fault in well *U-10k 1* (pl. 3).

The far western part of the northern alluvial–volcanic aquifer overlies or abuts the regional carbonate aquifer. In this area, water is assumed to flow westward from the alluvial–volcanic aquifer into the carbonate aquifer (pl. 3). Westward flow in this localized area is conceptualized because of an apparent hydraulic connection between the two aquifers and heads in the carbonate aquifer that are assumed to be lower than in the alluvial–volcanic aquifer. Hydraulic connections and heads in the western part of this alluvial–volcanic aquifer are poorly understood and flow directions, therefore, are uncertain.

Volcanic Confining Unit

The most noteworthy feature of the hydraulic heads in the volcanic confining unit is the elevated nature relative to heads in the alluvial–volcanic aquifer and the regional carbonate aquifer (pls. 1-4). Heads in the volcanic confining unit are as much as 100 ft higher than nearby heads in the alluvial–volcanic aquifer in the northern part of Yucca Flat and 10 to 30 ft higher further south. These elevated heads suggest that the volcanic confining unit is an important source of water to the adjacent aquifers, given no significant precipitation recharge or other lateral inflows to the alluvial–volcanic aquifer. Although an important source, the total flux into the aquifer is limited by the low hydraulic conductivity of the volcanic confining unit.

The heads in the volcanic confining unit were contoured with the sole purpose of conceptualizing flow from the volcanic confining unit into the alluvial–volcanic aquifer. The portrayal of the potentiometric surface in the volcanic confining unit, as shown on plate 3, is intended primarily to convey general concepts of flow in the unit rather than to precisely represent the true altitude of the potentiometric surface within the confining unit. The contoured steep horizontal hydraulic gradients, which range from about 50 to 300 ft/mi as compared to gradients in the alluvial–volcanic aquifer that range

from about 5 to 50 ft/mi, are consistent with a low hydraulic conductivity in the volcanic confining unit. Where not dissected by faults, the volcanic confining unit serves as a barrier to flow and transport between the alluvial–volcanic and carbonate aquifers.

The mapped potentiometric surface is, at best, an approximation because some of the heads posted on plate 3 could be influenced by potentially large, naturally occurring, vertical hydraulic gradients in the confining unit, and some heads may be misidentified as representative of predevelopment conditions. Heads could be misidentified because of the difficulty in identifying a representative head in a slow-equilibrating, low-permeability unit and some heads could be affected by underground nuclear tests. An attempt was made to remove nonstatic and test-affected heads, but uncertainty remains. The relatively large 40-ft contour interval is a reflection of the uncertainty in the mapped surface and the steeper hydraulic gradients in this unit. Where the alluvial–volcanic aquifer overlies the volcanic confining unit, the potentiometric surface of the top of the confining unit approximates the potentiometric contours for the aquifer (pl. 3).

In some cases, elevated heads mistakenly assigned as representative of predevelopment conditions and posted on plate 3 may actually be elevated because of nearby nuclear tests. However, enough non-impacted head measurements are elevated to provide a high level of confidence that the potentiometric surface in the volcanic confining unit is naturally elevated. These head measurements are unlikely to be affected by nuclear tests because no test was conducted near enough to these wells prior to the measurements to cause the heads to be elevated. Examples of wells with naturally elevated heads include *ER-6-1-2 piezometer*, *TW-E (1970 ft)*, *U-3mi*, *UE-4a (2655 ft)*, *UE-4ab (2396 ft)*, *UE-4av (1758 ft)*, and *WW-2 (2045 ft)* (pl.3, app. 1–3). These widely-spaced wells in the northern and eastern parts of Yucca Flat proper occur in areas where the volcanic confining unit occurs as the uppermost saturated zone.

The occurrence of elevated heads over a widespread area with virtually no recharge is problematic, but it is suggestive of a very slow-responding system. Elevated heads could occur because the hydraulic conductivity of the volcanic confining unit is so low that, even with no present-day recharge, heads have not yet equilibrated to post-pluvial conditions (Winograd and Thordarson, 1975). Higher past water tables have been suggested by various investigators (Levy, 1991; Marshall and others, 1993; Quade and others, 1995; D'Agnese and others, 1999). An alternative explanation for elevated heads is the possibility that minute amounts of present-day recharge are occurring, sufficient to maintain mounding in the volcanic confining unit. In low-lying areas of Yucca Flat, the water flux in the shallow unsaturated zone is upward (Levitt and Yucel, 2002; Walvoord, Plummer, and others, 2002; Kwicklis and others, 2006; National Security Technologies, 2007b), implying no infiltration from present-day precipitation. The exception to this may be small amounts of recharge focused along dry washes or in Yucca Lake playa during flooding events

(Doty and Rush, 1985; Stonestrom and others, 2003). However, focused recharge cannot explain the pervasive elevated heads measured in wells in the volcanic confining unit. A potential source of widespread recharge is slow infiltration of paleo-recharge derived from a wetter climate at the end of the Pleistocene about 10,000 years ago. An estimate of paleo-recharge to the water table in Yucca Flat at borehole UE-6e (fig. 2) is about 0.01 in/yr (Walvoord, Phillips, and others, 2002). Isotopic data from well *ER-2-1 main (shallow)*, open to the volcanic aquifer, suggest the water in this well was recharged under different climatic conditions (Farnham and others, 2006). This is consistent with the concept of a slow-draining alluvial–volcanic aquifer surrounded by a nearly impermeable volcanic confining unit. The source of the paleo-water could be either drainage of old water from the volcanic confining unit or current infiltration of paleo-recharge through the thick unsaturated zone.

The vertical-head distribution in the confining unit is unknown in areas where the confining unit is overlain by the alluvial–volcanic aquifer and underlain by regional carbonate aquifer. In these areas, no head measurements are available from wells completed solely within the volcanic confining unit. There are two possible alternatives to describe the vertical-head distribution in the volcanic confining unit where it lies between the two aquifers. One possibility is that the head profile in the confining unit has a relatively linear downward gradient transitioning from a higher head in the alluvial–volcanic aquifer to a lower head in the regional carbonate aquifer, indicating potential seepage of small amounts of groundwater into the underlying regional aquifer. Alternatively, the confining unit might have a higher head than in the bounding aquifers resulting in a groundwater divide within the confining unit. This alternative is supported by (1) the existence of previously discussed elevated heads elsewhere in the confining unit where recharge is expected to be negligible, (2) low hydraulic-conductivity estimates of the confining unit tuffs (1 x 10^{-5} ft/d) as demonstrated by extremely slow equilibration following well drilling or nuclear testing (Halford and others, 2005; Elliott and Fenelon, 2010), and (3) the existence of an elevated head about 10 mi south of the study area in Frenchman Flat (fig. 1) from a well that is open to a thick section of the volcanic confining unit sandwiched between aquifers with lower heads (see well *ER-5-4-2* in Elliott and Fenelon, 2010).

In either of the alternatives above, the confining unit is an effective barrier separating the alluvial–volcanic aquifer from the carbonate aquifer. Furthermore, the amount of water flowing through the volcanic confining unit is insignificant relative to flow in the aquifers. These concepts are supported by low hydraulic-conductivity estimates and relatively high horizontal hydraulic gradients for the volcanic confining unit, and by the presence of naturally occurring elevated heads in the alluvial–volcanic aquifer and volcanic confining unit in the absence of a significant recharge source.

Carbonate Aquifer System

The carbonate aquifer system consists of local and regional carbonate aquifers and the siliceous confining unit (fig. 3). The local carbonate aquifers drain into adjacent siliceous confining units, whereas the regional carbonate aquifer discharges to springs outside the study area. Potentiometric contours were constructed for the regional carbonate aquifer only (pl. 4). The local carbonate aquifer and the siliceous confining unit were not contoured because of limited hydraulic-head data. Additionally, the confining unit has steep hydraulic gradients that make it difficult to accurately contour the hydraulic-head distribution within this unit.

The highest carbonate heads in the study area occur in the local carbonate aquifers. Heads in these aquifers are elevated by more than 500 ft from heads in the underlying regional carbonate aquifer (pl. 4). This head difference is assumed to result from hydraulic isolation imposed by confining units that typically surround these local aquifers. Three local aquifers, two in western and one in northern Yucca Flat, have been pumped for local water supply or for scientific research directed at gaining a better understanding of radionuclide transport. The withdrawn water was pumped from wells completed in boreholes UE-2ce, UE-16d WW, and UE-15d WW (section *B–B'*, *C–C'*, and *H–H'*, pl. 1; pl. 4). Low to moderate water production from these wells (Elliott and Moreo, 2011) and their inferred hydraulic isolation support the classification of these small carbonate blocks as local aquifers. From a transport perspective, the local carbonate aquifer on the northwestern part of the study area and penetrated by borehole UE-2ce is notable (Carle and others, 2008; section *BB–BB'*, pl. 2; pl. 4). Near this borehole, an underground nuclear device was detonated about 200 ft above the water table in unsaturated carbonate rock, and eight other devices in the near vicinity were detonated in unsaturated tuff and alluvium overlying the carbonate rock (U.S. Department of Energy, 1997). Even under the most conservative assumption that radionuclides have entered or will enter this local carbonate aquifer, their transport into a more accessible downgradient environment would be severely hindered by the thick confining unit that hydraulically isolates this local aquifer from the regional carbonate aquifer.

The regional carbonate aquifer is subdivided into shallow and deep parts (fig. 11; pls. 1 and 4). The shallow part, represented by well data, is defined as the portion of the aquifer that is within about 6,000 ft of land surface (Fenelon and others, 2010). This report discusses flow only in the shallow regional carbonate aquifer. This focus on the shallow portion of the system is intended to provide information most pertinent to quantifying the hydraulic potential that controls the transport of radionuclides. Any transport would originate from the area around individual nuclear devices that were detonated in unsaturated rock or in the uppermost saturated zone at or near the water table. Once within the saturated zone, radionuclides likely will remain at relatively shallow depths as they are transported toward downgradient discharge areas. The deep parts of the aquifer are assumed to exert minimal influence on the transport of radionuclides off of the NNSS.

All but 7 of the 662 underground nuclear tests in the study area were conducted in the alluvial–volcanic aquifer system above the carbonate aquifer. Three tests were conducted in granite in the Climax Mine area. Four tests—Handcar, Kankakee, Bourbon, and Nash—were conducted in unsaturated carbonate rock (Carle and others, 2008). Two of the tests were conducted near each other in northern Yucca Flat and one test was conducted further south in NNSS Area 7. One of the northern tests, Kankakee, is shown on the northern end of section HH–HH' (pl. 2). The Bourbon test, conducted in Area 7 just northwest of borehole UE-7nS (pl. 4), had its device detonated within 150 ft of the water table and its cavity is predicted to intersect saturated carbonate rock at the top of the regional carbonate aquifer (Carle and others, 2008). A few of the tests in the alluvial–volcanic aquifer system were conducted in close proximity to the carbonate aquifer, such as Bilby (section CC–CC', pl. 2) and Strait (section GG–GG', pl. 2).

In contrast to the shallow regional carbonate aquifer, the deep part of the aquifer is assumed to be less active hydraulically, with low flow rates. Knowledge of flow in the deep part of the regional carbonate aquifer is limited because no well data exist. Deep regional flow likely originates as recharge in areas far upgradient of Yucca Flat, where it slowly migrates into the deep regional flow system. Once in the deep system, it travels long distances along regional flow paths to major areas of discharge, such as Ash Meadows or Death Valley. The interaction of this deep water with water in the shallow flow system is believed to be minimal (Tóth, 1962; Freeze and Witherspoon, 1967); as a result, analysis and interpretations are restricted to the shallow regional carbonate aquifer.

Flow paths in the shallow part of the regional carbonate aquifer are influenced by the deep part primarily in areas where the top of the aquifer is deeply buried and no shallow carbonate aquifer is present. In these areas, the regional carbonate aquifer is conceptualized to be so deep that adjacent groundwater in the shallow carbonate aquifer is preferentially directed toward other areas of shallow aquifer rather than moving into the deeper more stagnant part of the aquifer. The result is that the deep part of the aquifer and the overlying, less-permeable, confining-unit rock function as a flow boundary. This situation occurs along most of the western part of the study area, where only deep carbonate aquifer is present (pl. 4). This deep carbonate rock occurs in an area overlain by a thick wedge of siliciclastic rock (fig. 11; sections B–B' and C–C', pl. 1). The siliciclastic wedge and the resulting carbonate rock that is buried at great depth are conceptualized to separate shallow flow into a western and an eastern carbonate flow system, identified as the Shoshone Mountain and Yucca Flat tributary flow systems, respectively (Fenelon and others, 2010). The Shoshone Mountain tributary flow system forms the western boundary of the study area and is not discussed in this report (see shallow regional carbonate aquifer west of study area boundary on pl. 4).

Groundwater flow in the shallow regional carbonate aquifer in Yucca Flat is controlled primarily by the aquifer boundaries, areas of recharge and lateral inflows, structural features, and the heterogeneity and anisotropy of the aquifer. These properties of the aquifer system, in turn, control the hydraulic-head distribution, flow gradients, and flow rates through the system. The potentiometric surface of the shallow regional carbonate was contoured (pl. 4) on the basis of estimates of hydraulic heads from measurements of water levels open to the carbonate aquifer, in conjunction with known and inferred geology and a conceptualization of controlling factors on the flow system. Contours also were constrained along the study area boundary to be consistent with regional contours of the carbonate aquifer system from Fenelon and others (2010, pl. 4). The contours are used to determine flow directions, estimate hydraulic gradients, define interactions with other aquifer systems, and gain a more complete understanding of the flow system.

The shallow regional carbonate aquifer, which is present in most of the study area, is bounded by the siliceous confining unit forming the siliciclastic wedge on its western end, the siliceous confining unit from Climax stock on its northern end, and the siliceous confining unit forming the regional hydrologic basement on its northeastern end (pl. 4). To the southeast and south, the carbonate aquifer extends past the study area boundary (pl. 4). The shallow carbonate aquifer, in most areas, is underlain directly by deep carbonate aquifer. Exceptions are the northern part of the study area where the shallow carbonate aquifer directly overlies the siliceous confining unit (sections F–F', G–G', and H–H', pl. 1) and, locally, where thrusted regional carbonate aquifer is underlain by the siliceous confining unit (for example, western part of sections B–B', pl. 1). The top boundary of the shallow aquifer is, in most areas, the volcanic confining unit (fig. 10) or the water table where the volcanic confining unit is absent (pl. 1).

Water originating as recharge in highland areas internal (fig. 2) and external (fig.1; Hevesi and others, 2003; Fenelon and others, 2010) to the study area infiltrates directly into the carbonate aquifer or enters indirectly as groundwater flow through adjacent geologic units. Areas of direct infiltration to the carbonate aquifer are primarily on the eastern and southwestern parts of the study area. Surface exposures of carbonate rock on the eastern end of the study area in the highest parts of the Halfpint Range (pl. 4) promote rapid infiltration of precipitation to the carbonate aquifer. On the southwestern part of the study area, recharge in Mine Mountain and CP Hills (fig. 2) can infiltrate directly to the carbonate aquifer where the aquifer is exposed at the water table (fig. 7). Along the northern boundary and the remaining part of the western boundary, recharge in the highland areas (fig. 2) enters the water table in shallower geologic units and seeps into the shallow regional carbonate aquifer primarily along contacts with the siliceous confining unit (see seepage arrows along western, northern, and northeastern boundaries of carbonate aquifer on pl. 4). An example of this highland recharge is shown at well UE-2ce (section BB–BB', pl. 2), where the water composition is dominated by local recharge (Farnham and others, 2006). This recharge entering from the Eleana Range supports an elevated head in the local carbonate aquifer, which feeds water downward and eastward toward the regional carbonate aquifer. Inflows of groundwater occur across the southern half of the eastern study area boundary through carbonate rocks that extend outward to the east. Vertical gradients between contoured potentiometric surfaces in the alluvial–volcanic and the carbonate aquifers (pls. 3, 4) generally are downward and indicate local leakage to the carbonate aquifer from above, across the less-permeable volcanic confining unit that overlies the regional carbonate aquifer (pl. 2).

Hydraulic heads available for contouring the potentiometric surface of the regional carbonate aquifer are relatively sparse compared with heads for the alluvial–volcanic aquifer system. However, relative to carbonate heads available for the NNSS and surrounding region, Yucca Flat has a high density of hydraulic-head data for the regional carbonate aquifer (Fenelon and others, 2010). The carbonate-head data are located primarily in areas of nuclear testing and where the top of the carbonate aquifer is relatively shallow. Hydraulic-head measurements in wells open only to the regional carbonate aquifer range from 2,384 ft in well *WW-C (1373–1701 ft)* to 2,447 ft in well *ER-6-2*; these are the southernmost two carbonate wells in the study area. A few wells noted on plate 4 are open not only to the regional carbonate aquifer but also to overlying saturated volcanic confining unit or other non-carbonate rock. Three wells open to a composite of the regional carbonate aquifer and volcanic confining unit in the northern part of the study area have head estimates of slightly less than 2,500 ft (pl. 4). These composite heads are lower than the heads in the corresponding shallow wells open only to volcanic confining unit. For example, in borehole UE-10 ITS 5, the composite head is 2,483 ft (pl. 4), whereas the volcanic confining unit head is 2,535 ft (pl. 3). These composite heads were contoured as representative of the regional carbonate aquifer but could be elevated because of contributions of water from the volcanic confining unit. Hydraulic heads in some of the wells open to a composite of units, such as *UE-8e (2470 ft)* and *UE-4av (1724–2815 ft)*, clearly are elevated with respect to other nearby carbonate heads and are assumed to be influenced by groundwater conditions in non-carbonate rock. The dominance of head by non-carbonate rock may suggest that the relatively thin intervals of carbonate rock open to these wells are void of any major fractures. These elevated heads are considered anomalous and were not contoured on plate 4.

The predevelopment potentiometric surface in the shallow part of regional carbonate aquifer is defined by contours that range from 2,380 to 2,500 ft (pl. 4). These contours form an inverted V-shape pattern that roughly aligns with the north-south axis of Yucca Flat. Flow in the carbonate aquifer is characterized by a regional component of flow to the south along the axis of the "V", superimposed locally by inward flow from the east and west that drains toward the central axis.

Horizontal hydraulic gradients along the regional flow direction (north to south) are low, ranging from about 6 ft/mi on the far northern end of the basin to less than 2 ft/mi further south. The low north-south hydraulic gradient in the regional

carbonate aquifer is indicative of high aquifer permeability, very low flow rates, or a combination thereof. Gradients toward the central axis of the basin are higher, ranging from about 25–50 ft/mi in the northeastern part of the basin to 25 ft/mi in the western part. The gradients reflecting eastward and southwestward flow were calculated on the basis of the contours on plate 4 and are uncertain. The magnitudes of the eastward and southwestward gradients in the regional carbonate aquifer are comparable to typical horizontal hydraulic gradients in the alluvial–volcanic aquifer.

The pattern of contours in the regional carbonate aquifer can be explained, in part, by the shape of the boundary of the shallow carbonate aquifer with the siliceous confining unit (fig. 7; pl. 4). This physical boundary, in conjunction with inflows of recharge along the western and eastern boundary, causes the potentiometric contours to parallel the boundary.

The inverted V-shape pattern of the carbonate-aquifer contours, which generally parallels the carbonate aquifer boundary, is accentuated by the heterogeneity and anisotropy of the carbonate aquifer. The low-hydraulic-gradient, central corridor of Yucca Flat is parallel to the series of normal faults that transect the basin in a north-south direction (pl. 4). This low-gradient potentiometric trough, which extends southward to Ash Meadows (fig. 1), has been recognized by previous investigators and has been interpreted to be a highly transmissive corridor that is less than 3 mi wide (Winograd and Thordarson, 1975; Winograd and Pearson, 1976). This several-mile-wide trough in Yucca Flat is thought to be a zone of high transmissivity relative to the carbonate rock to the west and east. Additionally, the aquifer is highly anisotropic, with enhanced flow parallel to the trough as a result of the high degree of open faults and fractures along the axis of the trough. These concepts are supported by the predevelopment hydraulic heads and gradients in the carbonate aquifer (pl. 4), as well as a 90-day aquifer test at well *ER-6-1-2 main* (Stoller Navarro Joint Venture, 2005). The aquifer test demonstrated large and rapid responses directly north of the pumping well and muted or no responses to the east and west. Details of the aquifer-test results are discussed in the "Pumping" section later in this report. The expected result of a highly transmissive carbonate-rock corridor with enhanced permeability in a north-south direction is a low hydraulic gradient along the central axis of Yucca Flat and higher gradients to the west and east, where the aquifer transmissivity is lower.

The low-gradient potentiometric trough is exemplified by the supplemental 2,390-ft contour on plate 4. This contour encompasses an area within Yucca Flat of about 3 mi wide and 15 mi long. The hydraulic head in borehole ER-3-1, about 3 mi east of the 2,390-ft contour, is only 1 ft higher suggesting an extremely flat gradient between the trough and groundwater to the east.

Two alternative 2,380-ft contours are presented on plate 4. Both contours extend into north-central Yucca Flat in order to honor the low hydraulic-head measurements in wells *U-7a* and *U-3cn 5*. Note that the head for *U-3cn 5* may be 8 ft lower than posted on plate 4 if adjusted for the anomalously warm temperature in the well, as discussed in the "Analysis of Water Levels" section. Both contours also are drawn to imply the faults as major controls on flow by portraying narrow contours that parallel the faults. The primary difference between the alternative contours is on their southern ends. The eastern 2,380-ft alternative contour is based on a structural interpretation that assumes the potentiometric trough will follow the fault structures that control drainage out of Yucca Flat. As such, the southern end of this contour is portrayed to swing east to parallel the major structures in this area. The western 2,380-ft alternative contour is based on a geochemical interpretation that suggests a common source of water, based on similarities in water chemistry, for the carbonate well completions in boreholes ER-3-1 and WW-C (Farnham and others, 2006). If these two wells have a common source of water, the low in the potentiometric trough must occur west of WW-C. By constructing contours as shown on the western geochemical alternative, water is able to flow to borehole WW-C from the east rather than from the west. Potential concerns with the geochemical interpretation that implies westward flow include the following:

- The source of water to WW-C is likely to come from the north rather than from the east if anisotropy directs water to move in a predominantly southward direction parallel to fault structures.
- Contours must cross rather than parallel major fault structures in the southern end of Yucca Flat, contrary to the conceptualization that faults are a primary control on flow.
- The geochemical analysis (Farnham and others, 2006) is based primarily on WW-C samples that were collected after extensive long-term pumping. It's possible that pumping induced water into the well from a direction different from predevelopment conditions.
- Farnham and others (2006) demonstrated that it was possible (but less likely) to match the water chemistry in WW-C by mixing water from ER-6-2 rather than from ER-3-1.

In the western part of the study area but east of the CP thrust, thrusted carbonate aquifer overlies nonthrusted carbonate aquifer. The thrusted aquifer is underlain by (1) siliceous confining unit overlying deep carbonate aquifer, (2) siliceous confining unit overlying shallow carbonate aquifer, or (3) nonthrusted shallow carbonate aquifer with no intervening confining unit (pl. 4). These relations can be seen from east to west on section C–C' (pl. 1) between Carpetbag fault and the western end of CP thrust. In areas where thrusted carbonate aquifer directly overlies nonthrusted aquifer, the two aquifers are assumed to be in hydraulic connection and to function as a single aquifer. Where two shallow carbonate aquifers occur with intervening confining unit (see area surrounding well ER-6-2, sections E–E' and F–F', pl. 1), the potentiometric surface in each aquifer was contoured (pl. 4). Conceptually, the contours portray hydraulic-head distributions in the two carbonate aquifers to be similar. However, recharge in these areas is inferred to enter and elevate heads in the uppermost

carbonate aquifer and consequently the two sets of contours were constructed in a way that implies a downward vertical hydraulic gradient.

Water in the thrusted part of the shallow regional carbonate aquifer is conceptualized to flow primarily eastward to the hydraulic low in the center of the basin. As portrayed, the horizontal hydraulic gradient in the thrusted carbonate aquifer west of Carpetbag fault is relatively minor, but this portrayal is speculative. It also is possible that a steeper gradient occurs in the thrusted portion of the carbonate aquifer because of a poor hydraulic connection with the carbonate aquifer to the east, as portrayed in Fenelon and others (2010). Three wells in the thrusted aquifer provide hydraulic-head information that may be useful for determining the head distribution in the thrusted part of the aquifer. The southernmost well, ER-6-2, has a hydraulic-head estimate of 2,447 ft that is believed to accurately represent a predevelopment head (pl. 4, app. 2). This hydraulic head was contoured and suggests that, in the area of the well, the head in the thrusted carbonate aquifer is not significantly higher than heads to the east. Further north, well UE-1j has a hydraulic-head estimate of 2,507 ft (pl. 4). It is uncertain whether the estimate accurately reflects a predevelopment head (app. 2). Only one water-level measurement was made in this well, shortly after reaming and coring the borehole (Elliott and Fenelon, 2010). This head estimate was not contoured on plate 4 and is assumed to be anomalously high; however, it also is possible that the head estimate accurately reflects the predevelopment head in the thrusted carbonate aquifer at this well location. If it does, then this head estimate would suggest that a buildup of head occurs across Carpetbag fault. The third well that provides information about heads in the thrusted part of the carbonate aquifer is UE-2s, in the northern part of the thrusted section. An injection test done in this well produced a head measurement of 2,640 ft, but the head was still declining when the test was abandoned (app. 1). Thordarson and others (1967) stated that the static water level probably is below the bottom of the well, which is at an altitude of 2,613 ft. The head estimate for this well of less than 2,640 ft (pl. 4, app. 2) provides an upper-bound estimate of the hydraulic head in the carbonate aquifer at this location. The bottom of the thrusted carbonate aquifer at this well location is estimated to be at an altitude of 2,492 ft (app. 3), so it is possible that the aquifer is unsaturated beneath this well. The significance of the bounding head estimate in well UE-2s is that the hydraulic heads in the northern part of the thrusted carbonate aquifer can be no more than about 200 ft higher than heads east of Carpetbag fault. The contours on plate 4 portray heads that are about 80 ft higher than heads to the east, although these contour values are speculative. The contours are not intended to portray the exact head-distribution in this area, but rather to show the concept of eastward flow across Carpetbag fault toward the hydraulic low in central Yucca Flat.

Groundwater in the northeastern part of Yucca Flat is conceptualized to flow toward the center axis of the basin, similar to flow in the thrusted regional carbonate aquifer (pl. 4). Limited hydraulic-head data are available to support the contours in this area. Contours are interpreted to be steep relative to the center of the basin. Water is portrayed to slowly drain out of the siliceous confining unit that bounds the northeastern part of Yucca Flat and to flow toward the highly transmissive part of the carbonate aquifer. Extremely high heads in the confining unit, such as the head of 3,250 ft in well *UE-10aa* (pl. 4), support the concept that the confining unit is relatively impermeable. Similar to the area of thrusted carbonate aquifer, the contours in the northeastern part of Yucca Flat are intended to portray drainage toward the center of the basin rather than to indicate the exact values of the heads in this area.

The potential exists for a hydraulic connection between the regional carbonate aquifer in northern Yucca Flat and carbonate aquifer further north in Emigrant Valley (pl. 4). This is based on hydrostratigraphic framework models that portray a thin strip of continuous carbonate rock, east of borehole U-15k Test Hole, which connects Yucca Flat and Emigrant Valley (Faunt and others, 2004; Bechtel Nevada, 2006). The conceptualization presented here assumes no hydraulic connection is present (pl. 4), similar to an alternative hydrostratigraphic framework model ("hydrologic barrier in northern Yucca Flat") presented in Bechtel Nevada (2006). The large decrease in hydraulic head of about 2,000 ft between Emigrant Valley to the north and Yucca Flat (pl. 4) suggests that the Climax stock is part of a substantial hydrologic barrier at the northern end of Yucca Flat (Winograd and Thordarson, 1975; Fenelon and others, 2010). In addition to a high hydraulic gradient between these two areas, other indirect evidence contradicts any significant inflow from the carbonate aquifer in Emigrant Valley through this potential carbonate connection. This evidence includes (1) the nearby presence of Climax stock—an igneous granitic intrusive rock that has thermally altered the adjacent carbonate rock and decreased its hydraulic conductivity, (2) the geology at nearby borehole ER-8-1, which penetrated only saturated granitic rock (siliceous confining unit) and no saturated carbonate rock (section *G-G'*, pl. 1; pl. 4; app. 3), (3) no evidence in the regional carbonate aquifer in Yucca Flat of high-sulfate, high-chloride water typical of seeps from the granite in Climax Mine (Farnham and others, 2006); and (4) heads in wells *UE-15d WW (cased)*, *U-15k Test Hole*, and the *ME* wells near Climax Mine (pl. 4; app. 2), which indicate a consistent high gradient across the saturated rock that separates the two aquifers. Any potential hydraulic connection likely is small and is assumed negligible relative to total flow in the shallow regional carbonate aquifer within the study area.

Inflow to the regional carbonate aquifer in the Yucca Flat area from the northwest, north, and northeast is limited by the low permeability of the surrounding confining unit through which the majority of the inflow must pass. The inference of only limited lateral inflow from across these low permeability rocks is consistent with the steep hydraulic-head gradient found throughout their extent. The siliceous confining unit that occurs in the northeastern part of Yucca Flat restricts the amount of water coming into the study area from the northeast (Fenelon and others, 2010). Significant flow likely enters the study area from the east, south of borehole ER-3-1. Inflows

from the overlying alluvial–volcanic aquifer system, as well as from the west, north, and northeast, all converge to form a major southward flow path through Yucca Flat centered near Yucca fault (pl. 4). Although considered a major flow path for Yucca Flat, the amount of water moving beneath Yucca Flat is relatively minor, with estimates ranging from less than 350 acre-ft/yr (Winograd and Thordarson, 1975) to 1,000 acre-ft/yr (Harrill and other, 1988).

Water in the regional carbonate aquifer flows toward areas of progressively lower hydraulic head at the southern end of the study area and ultimately discharges at points southwest of the study area. The flow in Yucca Flat was mapped by Fenelon and others (2010) as part of the Yucca Flat tributary flow system, one of multiple groundwater tributary flow systems in the NNSS area that feed larger downgradient flow systems. The study area encompasses the upper two-thirds of the Yucca Flat tributary flow system. Flow paths in this tributary flow system, which extends to the southern end of the NNSS west of Mercury (fig. 1), are portrayed to split south of the study area. Water from the eastern flow path discharges into the downstream Ash Meadows flow system and water from the western flow path discharges into the Rock Valley tributary flow system (Fenelon and others, 2010, pl. 6). The part of the carbonate aquifer in the study area that discharges to each of these downgradient flow systems is imprecise because of limited data south of the study area. However, based on plate 6 of Fenelon and others (2010), it can be inferred that water derived from the western part of the study is more likely to discharge into the Rock Valley tributary flow system and water from the eastern part is more likely to discharge into the Ash Meadows flow system. The Rock Valley tributary flow system discharges water to the Alkali Flat–Furnace Creek Ranch flow system, which ultimately discharges water to the land surface southwest of the study area in southern Amargosa Desert and Death Valley (Fenelon and others, 2010). The Ash Meadows Flow system discharges water south-southwest of the study area to springs in Ash Meadows (fig. 1; Fenelon and others, 2010).

Transient Stresses

Transient water levels were identified and analyzed to provide information to better understand hydraulic responses to stresses and hydraulic connections within and between flow systems in Yucca Flat. The two primary anthropogenic stresses on the groundwater system since about 1950 are nuclear testing and pumping. All water levels measured in Yucca Flat through 2010 (app. 1) were examined to identify and flag those levels affected by these stresses. The effects of these two stresses on the groundwater system are discussed in the next two sections. A third stress on the groundwater system is natural recharge, which can cause both short- and long-term changes in water levels. Other transient stresses either are of short duration, such as daily and seasonal barometric changes, or are localized and isolated. An example of the latter situation is a rising water-level trend from 2004 to 2010 in well *WW-3 (1800 ft)* (app. 1) that is attributed to a localized mound resulting from artificial recharge (Elliott and Fenelon, 2010). The mound is theorized to be sourced by water from a nearby leaking pond infiltrating through the thick unsaturated zone.

Water levels affected by nuclear testing or pumping were not used to construct potentiometric surfaces representative of a predevelopment condition. A predevelopment condition assumes an equilibrium or near-equilibrium state in the groundwater flow system prior to any major changes that result from human activity. In the case of Yucca Flat, large water-level changes have been attributed to both pumping and nuclear testing. Even under natural predevelopment conditions, water levels can be in a state of dynamic equilibrium where levels fluctuate because of short-term and long-term changes in recharge. These fluctuations generally are small and can be ignored for the purposes of constructing generalized potentiometric surface maps. However, transient water-level fluctuations resulting from changes in recharge are examined to help conceptualize the flow systems in Yucca Flat.

Nuclear Testing

Nuclear tests detonated well above (greater than 328 ft or 100 m) the water table and near or below the water table are shown on figure 7. Grouping tests into these two categories was done originally by Bowen and others (2001) to identify tests that may contribute radionuclides directly to the regional water table and to quantify the radionuclide source term for transport modeling (Stoller-Navarro Joint Venture, 2009). This categorization is useful for understanding which tests may have released radionuclides directly into the water table but is less useful for determining which water levels may have been impacted by tests.

The magnitude and duration of a nuclear-test effect on a water level is related to the (1) yield of the test, (2) proximity of the test to the saturated zone, (3) distance between the test and the measured water level, and (4) properties of the geologic media present between the test and the measured water level. The interplay between these four variables can make it difficult to determine when and which water levels are affected by testing, especially where water-level data are limited and variables 1 and 4 are known only approximately. In general, however, the relative effect on a water-level measurement from a nuclear test is positively correlated with test yield and compressibility of the geologic media between the test and the measurement and negatively correlated with distance of the test to the saturated zone, distance between the test and the measurement, and hydraulic conductivity of the geologic media between the test and the measurement.

Test yields are reported imprecisely (U.S. Department of Energy, 2000b) for many of the tests and, as such, are approximations. Yields can be reported as an exact number (for example, 0.37 kilotons), a range or limit (for example,

20 to 200 kilotons), or a relative size (for example, "low"). In Yucca Flat, nearly one quarter of all tests were greater than 20 kilotons.

Nuclear devices in Yucca Flat were buried deep enough to prevent venting of radionuclides to the atmosphere (U.S. Congress, Office of Technology Assessment, 1989). Typically, smaller tests were buried greater than 300 ft below land surface and larger tests were buried greater than 1,000 ft. Because large-yield tests generally were buried deepest, nearly all tests conducted below the water table were greater than 20 kilotons (Pawloski and others, 2008).

During an underground nuclear explosion, high heat and pressure combined with a shock wave create a roughly spherical melt cavity that can range from about 10 to more than 500 ft in diameter (Pawloski and others, 2008; Stoller-Navarro Joint Venture, 2009). The test yield is the dominant factor controlling the size of the cavity, with the largest tests creating the largest cavities (Pawloski and others, 2008). The hydrologic and mechanical effects from the blast extend outward as a series of concentric zones that surround the cavity and include a crushed and pulverized zone, an intensely fractured zone, and an elastic-deformation zone; the aggregate extent of these zones may be six or more cavity radii (Laczniak and others, 1996). The rocks in this outer zone are highly compressed, which can create high pore-fluid pressures (Knox and others, 1965; Burkhard and Rambo, 1991). These high pore-fluid pressures dissipate quickly in aquifers and slowly in confining units (Halford and others, 2005). Within minutes to days of the test, rock above the cavity collapses into the cavity forming a chimney (Pawloski and others, 2008). For many of the tests, the collapse chimney extends all the way to land surface, in which case, a collapse sink is formed (Grasso, 2001).

The spatial relation between wells with water levels affected by nuclear tests (app. 1) and their proximity to water-level impact zones from the tests is shown on figure 12. The impact zone is defined by the extent of a spherical zone that extends outward six cavity radii from the center of the detonation and occurs below the water table. As a result, all devices detonated below the water table and only those unsaturated tests occurring within six cavity radii of the water table are included. Many of the spherical zones have similar radii because they were based on categorized reported test yields (U.S. Department of Energy, 2000b). For example, all tests below the water table with a yield of "20 to 200 kilotons" have similar six-cavity radii (1,260 to 1,380 ft) calculated using the maximum reported yield of 200 kilotons. The impact zones are intended to show a relative area of influence for water levels that may be impacted by a nuclear test. The choice of six cavity radii is somewhat arbitrary as the impact area has been suggested to extend out from the center of the cavity to a distance of 2 to 4 cavity radii (Wohletz and others, 1999), about 6 cavity radii (Laczniak and others, 1996), and about 20 cavity radii (Tompson, 2008). A shortcoming of figure 12 is that it shows all tests conducted in Yucca Flat but omits the temporal relation between tests and water-level measurements. For example, a water-level measurement in appendix 1 that was

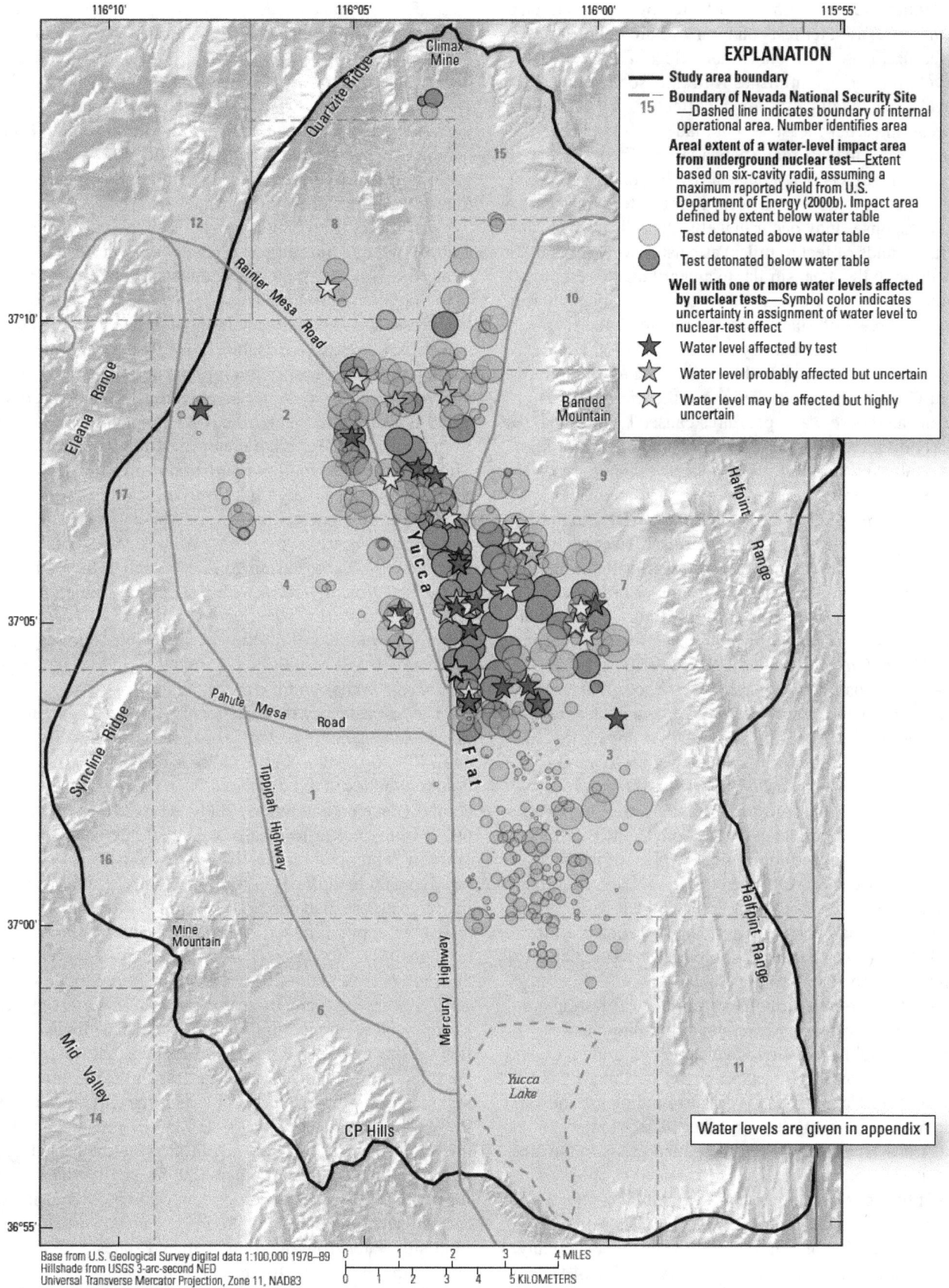

Figure 12. Water-level impact areas from underground nuclear tests and their spatial relation to wells having one or more water levels that may be affected by nuclear tests in Yucca Flat, Nevada.

made in January 1968 can only be affected by tests conducted prior to this date; all later tests on figure 12 are not relevant to this measurement. Additionally, the effect of a test on a water-level time series diminishes with time so that a recent water level from a well may show no effect from a particular test whereas an earlier measurement made in the same well immediately after the test may show a large effect.

Many of the effects that a nuclear test has on a water level are brief (seconds to days) and occur immediately following a test. Brief responses may occur in wells large distances from the test (thousands of feet to miles) as seismic waves from the test pass through the area, similar to an earthquake response. An example of this type of response from the Miniata nuclear test in NNSS Area 2 of Yucca Flat was recorded 46 mi to the south in Devils Hole (Dudley and Larson, 1976, fig. 6). Responses also occur in permeable units, such as alluvium, welded tuff, and carbonate rock, but are short in duration because elevated pore-water pressures caused by a nuclear test can quickly dissipate as water flows away from areas of high pressure (Halford and others, 2005). Brief responses typically were not recorded in water-level measurements because most wells were monitored infrequently (quarterly, annually, or sporadic) or only were monitored regularly (weekly to monthly) for a year or less after the well was drilled. Continuous water-level monitoring was uncommon prior to the end of testing in 1992. Some examples of short-term responses that are not routinely observed in the monitoring record occur in wells *TW-E (2620 ft)*, *U-2dr*, and *U-3cn 4 HTH* (app. 1).

Fifty-eight wells from 42 boreholes in Yucca Flat had at least one water level that may be affected by a nuclear test ("TR nuclear flag" = "Yes", "Yes?", or "?" in app. 1; fig. 13). All test-affected wells are located in the central to north-central part of the study area (fig. 13). This is to be expected given that this part of the study area contains most of the tests with potentially large effects in the saturated zone (fig. 12) and most of the wells analyzed for this study. Most of the water-level responses occur in wells open to the volcanic confining unit, a unit where only about 20 percent of the tests in Yucca Flat were conducted, but where more than 50 percent of the large-yield tests (greater than 20 kilotons) were conducted (fig. 13; Stoller Navarro Joint Venture, 2009).

Sustained water-level responses from nuclear testing are grouped into three categories: test-cavity infilling, depressurization of a highly pressurized volcanic confining unit, and miscellaneous factors (fig. 14). Test-cavity infilling (fig. 14*A*) results from filling a test cavity with surrounding groundwater following cavity creation after a nuclear blast. Moments after the blast, water levels are lowered instantaneously in the area of the cavity as water is expelled or vaporized by the blast. The rate of filling is dependent on the hydraulic conductivity of the materials surrounding the cavity. For tests detonated in the volcanic confining unit, the water level can take many years to rise and again approach pretest levels (fig. 14*A*). However, this effect generally is localized to the area of the cavity and has only been observed in wells completed in the cavity. Well *U-4u PS 2A* was drilled south of the Dalhart nuclear test

(fig. 15) and slanted northward to intersect the test cavity at depth. A rise of about 150 ft was recorded in this well over a 7-year period (fig. 14*A*) before measurements were discontinued. At the time of the last measurement, the water level still was rising and already was elevated more than 100 ft above the assumed level at the test location prior to testing. From here, it is expected that water levels will slowly decline as water in the surrounding volcanic confining unit depressurizes to reach equilibrium with pretest levels.

The effects from depressurization of the volcanic confining unit following a nuclear test typically show a steady declining trend (fig. 14*B*). Water levels decline as water flows from areas of high hydraulic head to areas of low head. In an intervening, pressurized volcanic confining unit, the flow of water is upward toward the water table, downward toward the regional carbonate aquifer, and lateral into the test cavity away from the pressurized zone. Water levels can be elevated more than 1,000 ft near the test and, based on extrapolation of trends from figure 14, the decline may persist for 100 years or more. This extreme pore-fluid pressurization and slow depressurization only occurs in very tight (low permeability) materials, such as the volcanic confining unit where hydraulic conductivity estimates are on the order of 1 x 10^{-5} ft/d (Halford and others, 2005). Large sustained effects such as those shown on figure 14*B* typically occur within six cavity radii of a nuclear test (wells *UE-4t 1 (1906–2010 ft)* and *UE-4t 2 (1564–1754 ft)* on fig. 15). In areas of multiple tests, the pressure response can be amplified as a result of overlapping effects (well *ER-2-1 piezometer (deep)* on fig. 15).

The last category of water-level response to a nuclear test, miscellaneous factors, includes examples from wells *UE-2ce* and *TW-7* (fig. 14*C*). Well *UE-2ce* is open to a thrust block of dolomite believed to be isolated from the regional carbonate aquifer (Fenelon and others, 2008). The well was drilled about 600 ft from nuclear test Nash, which was detonated in 1967. Water was pumped from well *UE-2ce* from 1977 to 1984, and briefly again in 2008. This well could only sustain pumping rates of less than about 10 gal/min (Elliott and Fenelon, 2010). After pumping ended in 1984, water levels recovered through 1994. "Recovered" water levels in 1994 were almost 50 ft lower than the first water level measured in the well in 1977. Following recovery from pumping, water levels began a steady decline of about 6 ft through 2011. The likely explanation for this extended long-term decline is that water levels are affected by the nearby nuclear test Nash. It is interpreted that the prepumped water level in *UE-2ce* was elevated by the Nash nuclear test 10 years earlier. The process by which the water levels became elevated following the Nash test is documented in Carle and others (2008). They propose that the formation of the nuclear-test chimney and the fracturing of rock adjacent to the test resulted in enhanced vertical drainage of in-situ water from overlying perched zones. This water created a mound of up to 100 ft in the chimney, which spread out to include the area around well *UE-2ce*. Current water levels suggest that the proposed mound is still dissipating more than 40 years after the test.

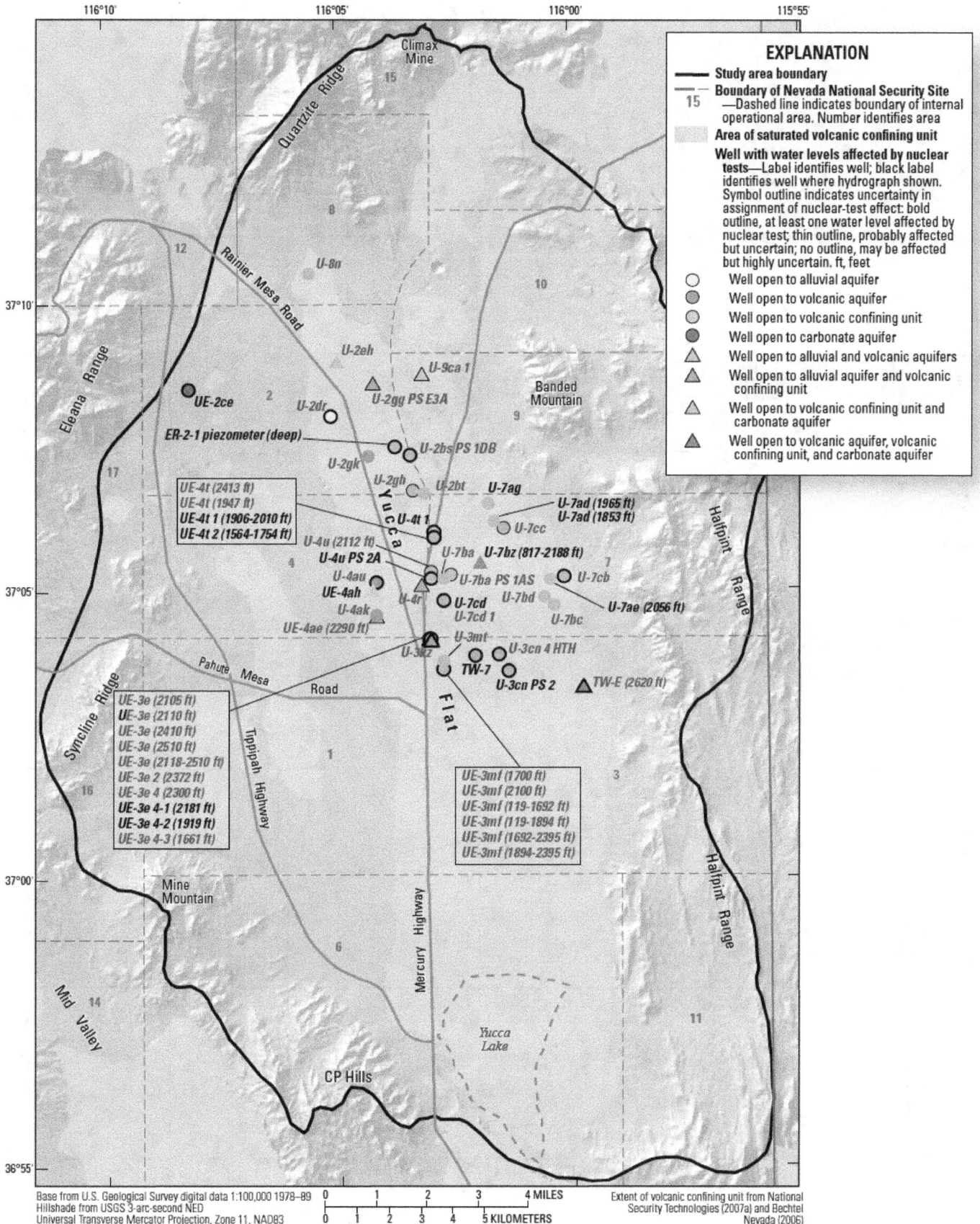

Figure 13.   Wells having at least one water level potentially affected by nuclear tests in Yucca Flat, Nevada.

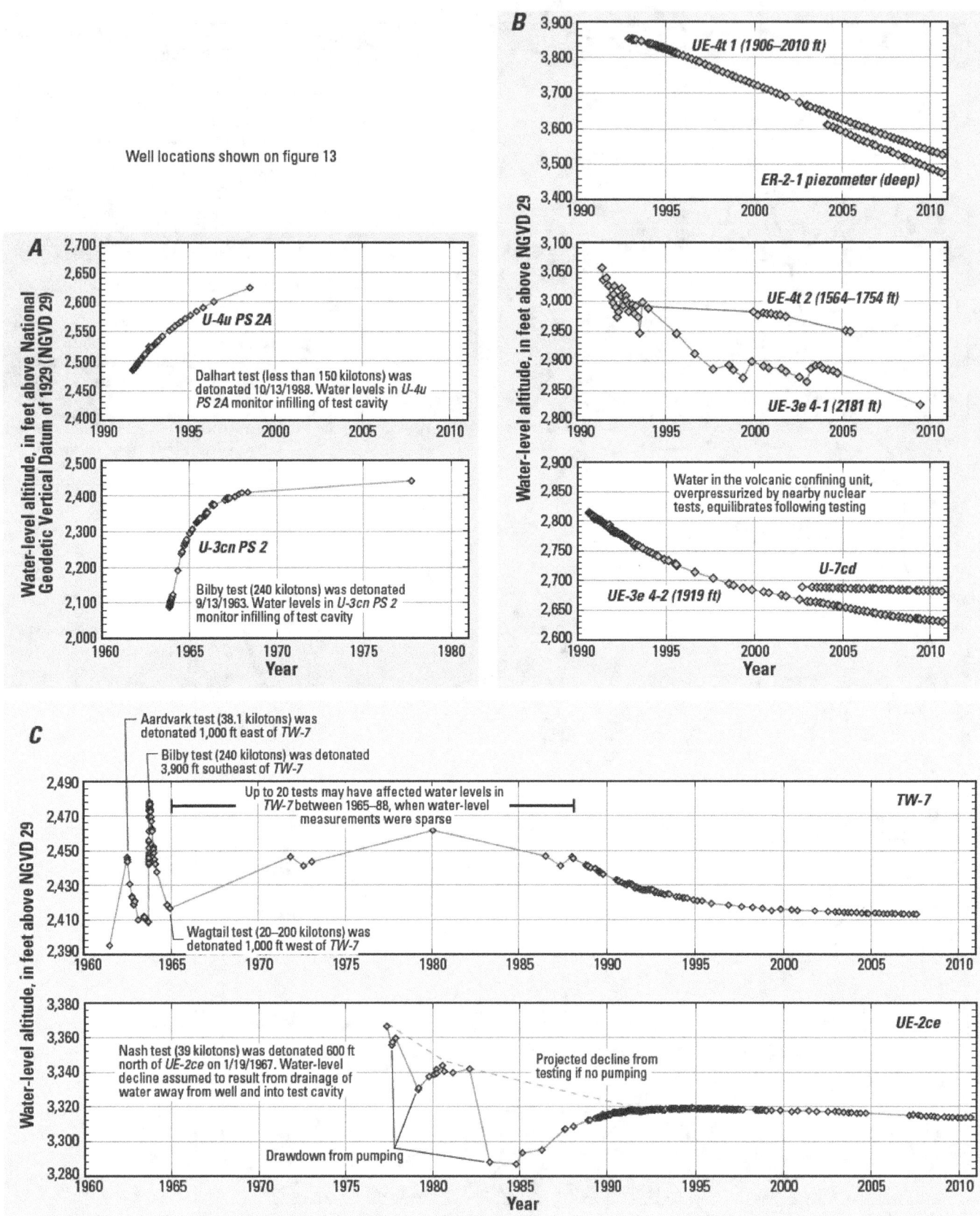

Figure 14.   Hydrographs of selected wells showing long-term effects from nuclear testing: **A**, test-cavity infilling, **B**, depressurization of a highly pressurized volcanic confining unit, and **C**, miscellaneous factors.

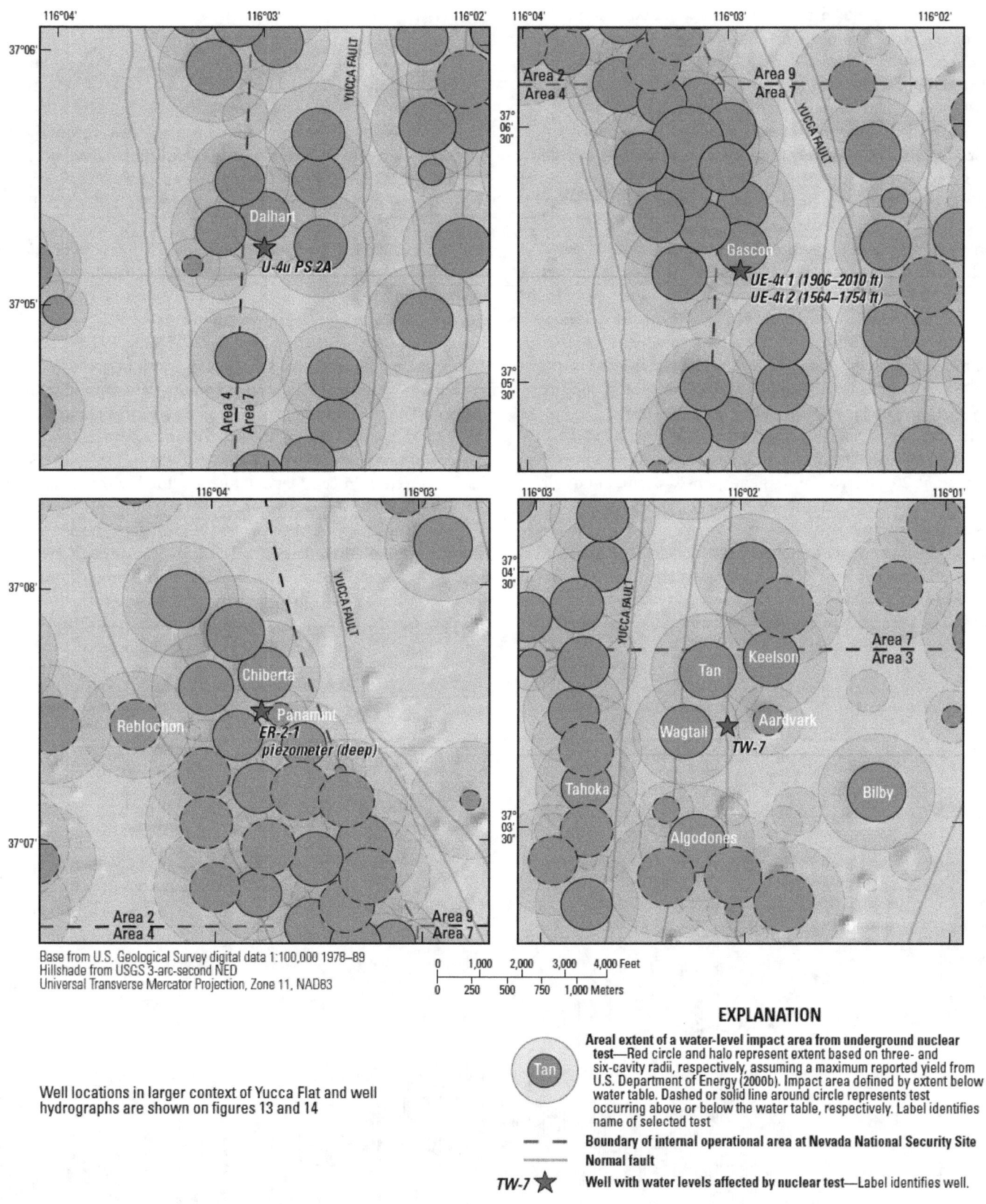

Base from U.S. Geological Survey digital data 1:100,000 1978–89
Hillshade from USGS 3-arc-second NED
Universal Transverse Mercator Projection, Zone 11, NAD83

0   1,000   2,000   3,000   4,000 Feet

0   250   500   750   1,000 Meters

EXPLANATION

Areal extent of a water-level impact area from underground nuclear
test—Red circle and halo represent extent based on three- and
six-cavity radii, respectively, assuming a maximum reported yield from
U.S. Department of Energy (2000b). Impact area defined by extent below
water table. Dashed or solid line around circle represents test
occurring above or below the water table, respectively. Label identifies
name of selected test

— — Boundary of internal operational area at Nevada National Security Site

Normal fault

TW-7 ★ Well with water levels affected by nuclear test—Label identifies well.

Well locations in larger context of Yucca Flat and well
hydrographs are shown on figures 13 and 14

Figure 15.   Spatial relations between water-level impact areas from underground nuclear tests and four wells having water levels that
show transient effects from nearby nuclear tests, Yucca Flat, Nevada.

Well *TW-7* is the best example in Yucca Flat of large responses from multiple and somewhat distant tests (fig. 14*C*). Clear responses from the Aardvark and Bilby tests were recorded in the water-level record. The Bilby test is about 3,900 ft, or 17 cavity radii, southeast of well *TW-7*, whereas the Aardvark test was closer but detonated in the unsaturated zone (fig. 15). Water levels in this well demonstrate that under certain conditions, the impact of a test on a water level can be large at relatively large distances from a test. The water-level effect from the Bilby test had nearly dissipated in *TW-7* a year and a half after the detonation. However, other closer tests such as the Wagtail test, detonated on March 3, 1965, about five cavity radii away (fig. 15), probably had a larger and more sustained effect on water levels in *TW-7*. Nonetheless, no water-level measurements were made for many years after the test to confirm a response (fig. 14*C*). The water-level response in *TW-7* was successfully simulated by Tompson (2008) as pore-pressure changes from multiple nearby tests.

In summary, water-level responses can be large and sustained in close proximity to nuclear tests. At distances of as much as 20 cavity radii, large responses occur, although they are less likely to be sustained for long periods of time. Sustained responses only occur in low-permeability materials such as the volcanic confining unit. Water-level trends in wells where pore waters have been pressurized are downward; rising trends are likely only in the cavity and surrounding area where water fills the cavity after water was vaporized and expelled following the detonation.

Pumping

Water has been withdrawn from aquifers in Yucca Flat since 1952. Through 2010, approximately 3,300, 3,700, and 11,000 acre-ft of water have been pumped from alluvial, local carbonate, and regional carbonate aquifers, respectively (fig. 16). Withdrawals from the volcanic aquifer have been negligible. Well *UE-1r WW* is completed across the volcanic and regional carbonate aquifers, but only about 70 acre-ft of water has been withdrawn from this well (U.S. Geological Survey, 2011; Elliott and Moreo, 2011). Patterns in total annual water withdrawals correlate with nuclear-testing activities in Yucca Flat. Only a few nuclear tests occurred prior to 1962, and pumping was minor. More than half of the Yucca Flat tests were conducted from 1962 to 1970 (U.S. Department of Energy, 2000b), which was a period of large withdrawals (fig. 16). Combined withdrawals from alluvial, local carbonate, and regional carbonate aquifers peaked in 1969 at 860 acre-ft. From 1971 to 1991, rates of testing were lower but remained relatively constant. Since 1991, no testing has occurred in Yucca Flat and average total withdrawals from 1991 to 2010 were about 170 acre-ft/yr.

Two wells, *WW-3 (1800 ft)* and *WW-A* (fig. 2), supplied water from the alluvial aquifer in Yucca Flat through 1988 (fig. 16). Since 1988, no water has been withdrawn from this aquifer in Yucca Flat (fig. 16). The alluvial aquifer generally is unconfined and has a large porosity compared to fractured bedrock aquifers. These hydraulic characteristics are

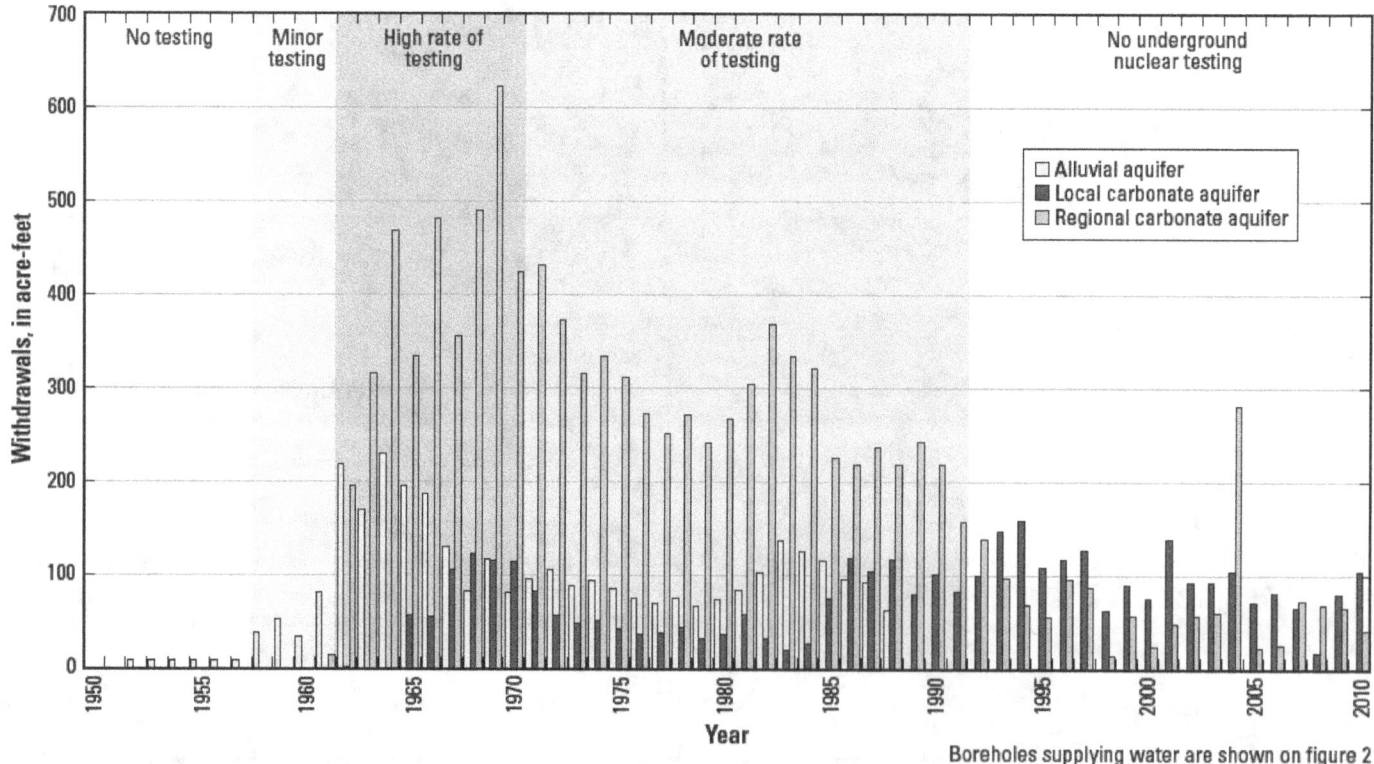

Boreholes supplying water are shown on figure 2

Figure 16.   Groundwater withdrawals from aquifers in Yucca Flat, Nevada, and their relation to rates of underground nuclear testing, 1950–2010.

consistent with the alluvial aquifer producing large amounts of water from a relatively small aquifer volume. The result is a deep, localized cone of depression around the pumping well, as occurred at well *WW-3 (1800 ft)* in the southern part of Yucca Flat. This well was pumped from 1952 to 1970. Shortly before pumping ceased, drawdowns exceeded 60 ft (fig. 17). About a mile away at well *UE-6d*, drawdown measured about 9 ft. Recovery to equilibrium conditions following pumping in the alluvial aquifer was long. In the case of well *WW-3 (1800 ft)*, full recovery took about 25 years from the time of maximum measured drawdown in 1969, whereas water levels in *UE-6d* were still rising in 2010 (fig. 17). Because of the localized nature of drawdown in the alluvial aquifer and the limited number of wells pumping from this aquifer, only a few water levels measured in the alluvial and volcanic aquifers were affected by pumping.

Withdrawals from local carbonate aquifers have been relatively consistent at a rate of about 50–100 acre-ft/yr from 1965 to 2010 (fig. 16). More than 99 percent of the production was from two wells completed in boreholes UE-15d WW and UE-16d WW (fig. 2). UE-15d WW pumped water from a deep dolomite aquifer within Precambrian rocks (Elliott and Moreo, 2011). The aquifer is portrayed in this report as isolated from the shallower regional carbonate aquifer. Water production from this well was limited and pumping ended in 1981 (Elliott and Moreo, 2011). UE-16d WW produces water from a Pennsylvanian limestone aquifer located in the vicinity of Syncline Ridge (fig. 2). Seventy percent of the production from local carbonate aquifers in Yucca Flat was from this well (Elliott and Moreo, 2011). Pumping from this well began in 1981 and continues today (Elliott and Moreo, 2011). The local carbonate aquifer is, by definition, isolated. Therefore, any large-scale production is limited, and although drawdowns can be relatively large at the pumping well and within the aquifer, they don't propagate large distances across Yucca Flat.

More than 60 percent of the water withdrawn from Yucca Flat was derived from the regional carbonate aquifer (fig. 16). Nearly all of the water was produced from three production wells: *WW-2 (3422 ft)*, *WW-C (recompleted)*, and *WW-C-1* (figs. 18, 19). Of these wells, only *WW-C-1* has produced water since 1995 (Elliott and Moreo, 2011), and withdrawal rates have been low.

A 90-day aquifer test in 2004 at well *ER-6-1-2 main* demonstrates the high diffusivity[1] of the confined regional carbonate aquifer in a north-south direction along the east side of Yucca Flat. Responses were rapid and extensive to the north of the pumping well. At well *ER-7-1*, more than 6 mi north of the pumping well, a response was noted almost immediately (within hours) after pumping began (Stoller-Navarro Joint Venture, 2005). Furthermore, the magnitude of the drawdown at well *ER-7-1* was about 60 percent of the drawdowns observed at wells *ER-6-1-1* and *ER-6-1 main*, about 200 ft from the pumping well (fig. 18). Drawdowns

---

[1] Diffusivity is a hydraulic parameter defined as the ratio of aquifer transmissivity to storage coefficient and is indicative of an aquifer's ability to transmit a pressure response due to a stress.

were less dramatic (*ER-3-1-2*) or unobserved (*UE-1h*) to the east and west of the main fault-controlled corridor of Yucca Flat (fig. 18). The aquifer test demonstrates how a stress can quickly propagate large distances through parts of the regional carbonate aquifer system in Yucca Flat. The propagation of the drawdown response is enhanced in a north-south direction parallel to the major normal faults. These faults, and the associated ancillary faults and fractures, are assumed to create a preferentially enhanced zone of groundwater flow through the center of the basin.

The aquifer test at *ER-6-1-2 main* (fig. 18) provides insight into the likely response of water levels in the regional carbonate aquifer to long-term pumping in the aquifer (fig. 16). Water-level measurements east of Carpetbag fault may be affected by current and historic pumping in the regional carbonate aquifer. Maximum responses likely are small (no more than several feet) and most of the response would dissipate quickly once the pumping ceases. An example of potential responses in well *UE-7nS*, open to the regional carbonate aquifer, to pumping from the aquifer is shown on figure 19. A clear response to pumping is seen in 2004 during the multi-well aquifer test at well *ER-6-1-2 main*. Less clear are the causes of other long-term water-levels fluctuations in *UE-7nS*. These fluctuations likely are a combination of responses to recharge, responses to varying amounts of pumping from the regional carbonate aquifer, and imprecise measurements prior to 1996 resulting in measurement errors of up to 1 ft. The combined long-term fluctuations from pumping and recharge generally are less than 5 ft and are unlikely to alter interpretation of predevelopment potentiometric conditions.

Recharge

Recharge in Yucca Flat is limited by low precipitation rates and high potential evapotranspiration rates. Most recharge likely occurs only during the winter and spring of very wet winters. During these cold, wet periods, rain and snowmelt can saturate soils and drive soil water downward past the zone of evapotranspiration where it can become recharge. Recharge is more likely to occur (1) in highland areas where precipitation is greater and snow can accumulate, (2) over permeable soils and bedrock where water can quickly move downward below the root zone, and (3) in areas where runoff can accumulate, such as in stream channels and playas (Flint and others, 2004). Furthermore, decadal-scale climatic cycles and El Nino years influence recharge potential. For example, large annual recharge events, especially during El Nino years, were more likely to occur from 1976 to 1999 than from 1956 to 1975 (Flint and others, 2004).

Areas of potential groundwater recharge, modified from Fenelon and others (2010), are shown on figure 2 and plate 3. The areas of recharge were generalized to reflect areas where land-surface altitude was greater than 6,000 ft and simulated net infiltration from Hevesi and others (2003) exceeded about 0.1 in/yr. With the exception of the highland areas rimming the study area, most of the study area is estimated to have no

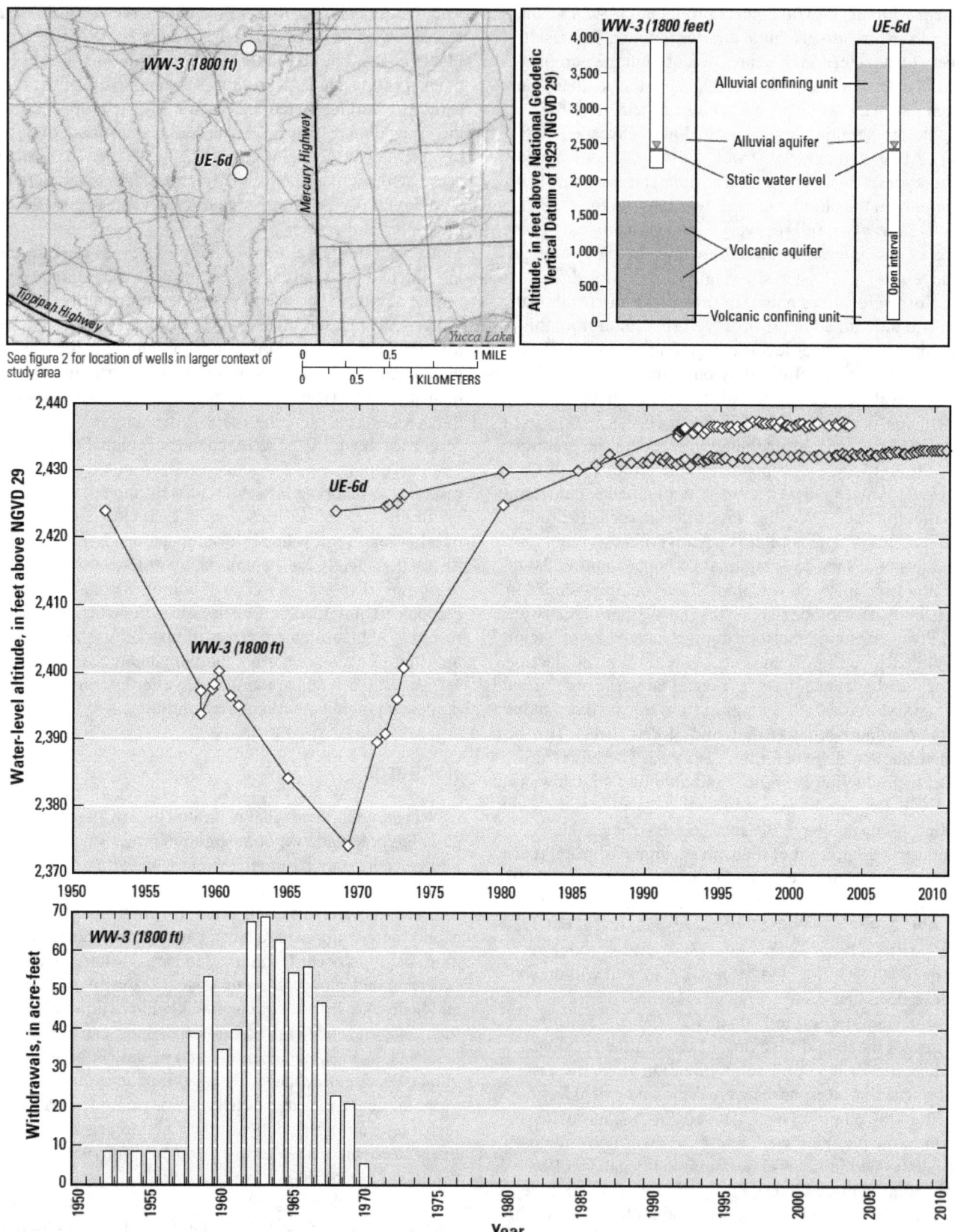

Figure 17. Relation between water levels in wells *WW-3 (1800 ft)* and *UE-6d*, completed in the alluvial aquifer, and withdrawals from well *WW-3 (1800 ft)*, Yucca Flat, Nevada, 1950–2010.

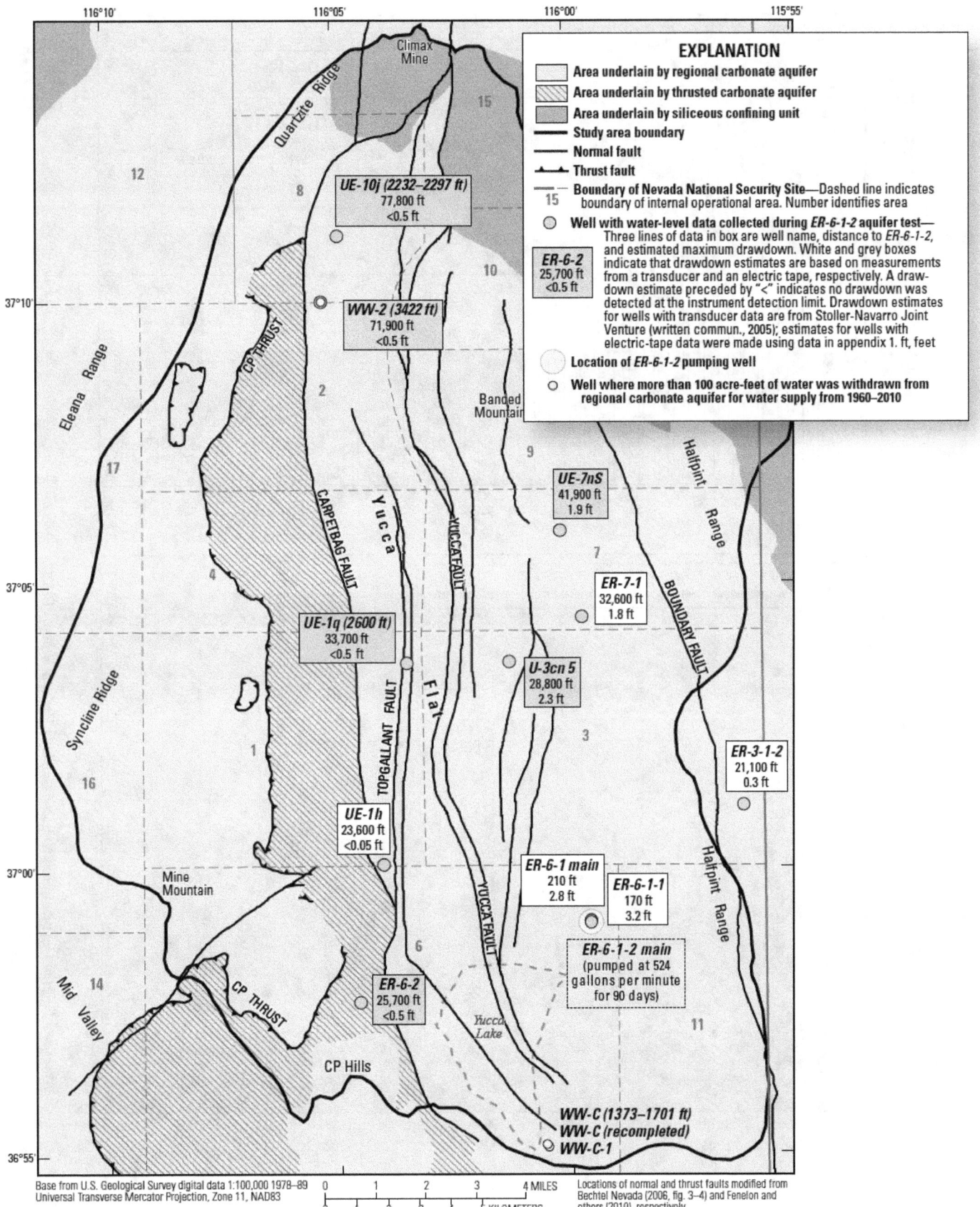

Figure 18. Observation-well responses from pumping the regional carbonate aquifer during a 90-day aquifer test at well *ER-6-1-2 main* in Yucca Flat, Nevada.

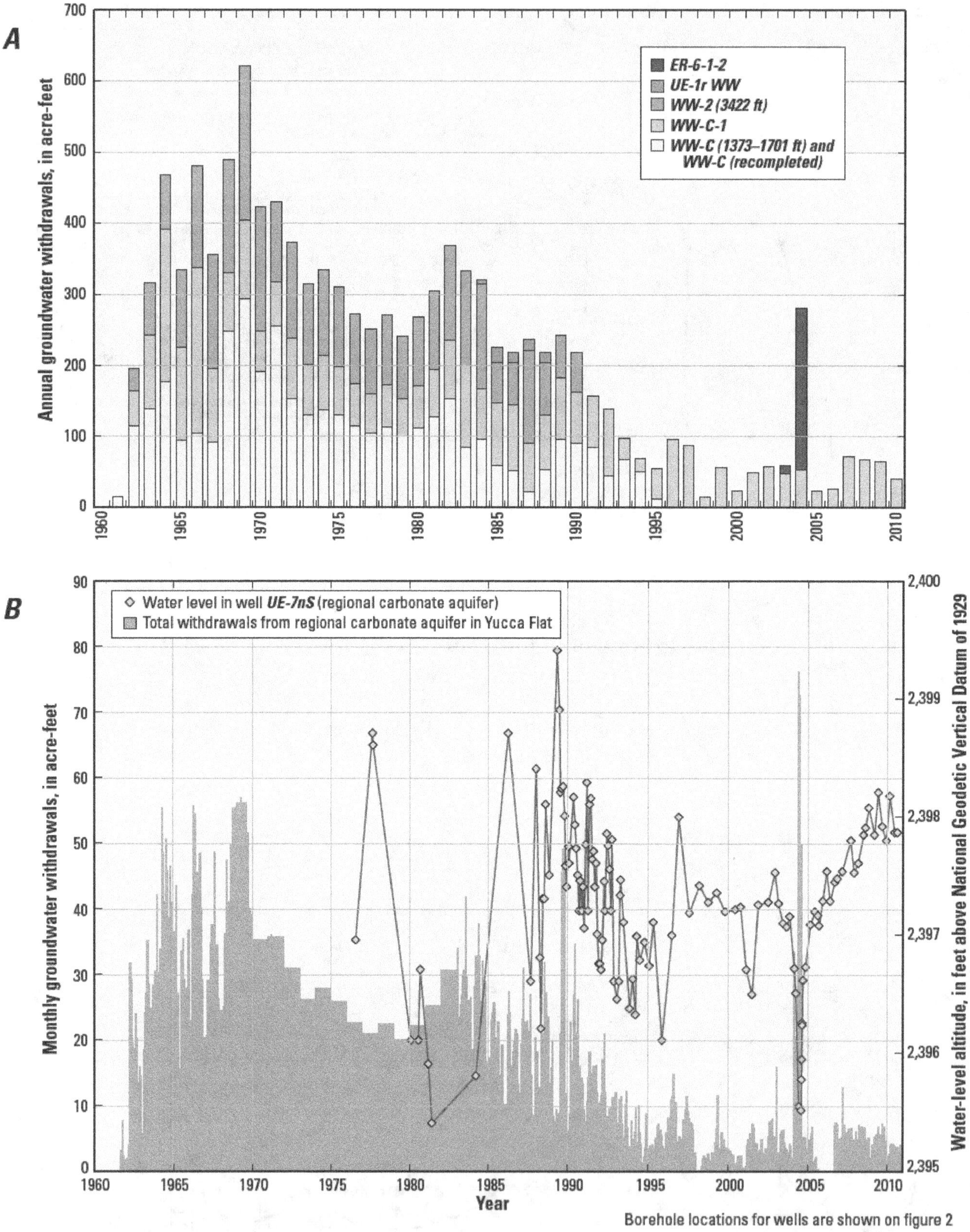

Figure 19. *A*, Annual withdrawals, by well, from the regional carbonate aquifer in Yucca Flat, Nevada, 1960–2010, and *B*, relation between total monthly withdrawals and water levels in well *UE-7nS*.

net infiltration (Hevesi and others, 2003). Areas in the study area estimated to exceed 0.2 in/yr are Syncline Ridge, Mine Mountain, and CP Hills, where infiltration rates are estimated to be as high as 0.8 in/yr (Hevesi and others, 2003).

Many water levels in Yucca Flat show naturally occurring transient effects that result from short-term and long-term variations in recharge rates. Long-term (50 years) changes in water levels resulting from recharge generally are less than 5 ft. Water levels are assumed to fluctuate around a long-term mean in a state of dynamic equilibrium. For the purposes of developing potentiometric surfaces representing predevelopment conditions, these natural fluctuations are assumed negligible relative to most vertical and horizontal gradients and can be ignored. However, analysis of water-level trends between wells can provide insight into areas that respond quickly or slowly to recharge events and aquifers that may be connected or isolated from each other.

Water levels from 16 wells with hydrograph records greater than 15 years are shown on figure 20. Water-level trends from these wells are attributed predominantly to changes in natural recharge, with little or no effect from pumping, nuclear testing, or other non-natural causes. Most of the water levels are from wells open to the carbonate aquifer or the siliceous confining unit. Most of these wells lie outside the central part of Yucca Flat (fig. 21). The lack of suitable long-term hydrographs showing natural fluctuations in central Yucca Flat is because most available hydrographs representative of alluvial and volcanic aquifers in this area are affected by nuclear testing or pumping.

Water-level measurements on figure 20 are shifted to arbitrary datums in order to facilitate comparison of the magnitude and direction of changes between hydrographs. Additionally, a LOcally WEighted Scatterplot Smooth (LOWESS) line was fitted to the water-level data to help identify trends. LOWESS is a nonparametric method of fitting a curved line to data (Helsel and Hirsch, 1992). At each data point, a predicted value is computed using a weighted linear regression. Predicted values are then connected to create a smoothed line. The line can be useful for discerning a pattern or trend in data with scatter.

LOWESS trend lines were grouped together by similar trends from 1987 to 2011 (fig. 22). Prior to grouping similar trends, data points for each LOWESS line were normalized between 0 and 1 for the period 1995–2011, a period where most wells on figure 20 were being measured at a regular frequency. Water levels prior to 1995 were normalized relative to the period 1995–2011. By doing this, normalized values of less than 0 and greater than 1 could occur if the water levels prior to 1995 fell outside the range of water levels from 1995 to 2011. Normalizing the data allowed for comparison of trends that were masked by large differences in the magnitude of water-level changes between hydrographs.

The largest water-level changes occurred in the western part of the study area, west of Yucca fault, in trend groups *A* and *B* (figs. 20, 21). Most large changes were measured in the carbonate aquifer in wells *ER-6-2*, *UE-1h*, *WW-2 (3422 ft)*, and *UE-10j (2232–2297 ft)*. In general, the smallest changes

were measured in the siliceous confining unit in wells *UE-1a*, *UE-1b*, and *UE-16f (1479 ft)* and in the volcanic aquifer in wells *TW-B*, *UE-1q (2600 ft)*, and *UE-1c*. Most of these wells are in trend groups *C* and *D* (figs. 20, 21). The magnitude of change in these wells likely is a function of the proximity of the measured well to a recharge area and the diffusivity of the water-bearing units between the well and the recharge area. Recharge from infiltration of precipitation occurs predominantly in highlands on the western side of the study area (fig. 21). Water-bearing units with high diffusivity typically are fractured and confined, such as the carbonate aquifer in Yucca Flat. This high diffusivity enables the aquifer to transmit a pressure response from a recharge pulse rapidly and over long distances, as was demonstrated with a pumping pulse during the *ER-6 1-2 main* aquifer test (fig. 18). The volcanic aquifer, which occurs primarily in the central part of the Yucca Flat basin, commonly is unconfined and is isolated from recharge laterally and vertically by confining units (fig. 7; plate 2). As such, water-level responses in the volcanic aquifer from recharge are expected to be minimal.

Five distinct trends were observed in the water-level data from 1987 to 2011 in Yucca Flat (fig. 22). Wells with hydrographs representing these trend groupings occur in distinct parts of the study area (fig. 21). Water-level trends in these grouped areas probably are controlled primarily by local recharge patterns, which are influenced by the water-bearing units that occur between the recharge source and the open interval of the well. Differences in recharge patterns may partly result from different recharge lag times through varying thicknesses of unsaturated zone. Recharge is shown qualitatively on fig. 22 as winter periods (October–March) when precipitation was extreme, based on the average of three long-term (1965–2010) precipitation monitoring stations. These precipitation stations (MV, A12, and PHS; Air Resources Laboratory, Special Operations and Research Division, 2011) are shown on figure 1 and are in the southwestern, northwestern, and northeastern parts of the study area. Winter precipitation is used to indicate years with potentially significant recharge because most recharge is derived from precipitation during this period (Winograd and others, 1998). Winters in the study area with extreme precipitation (1993, 1995, 1998, and 2005; fig. 22) had nearly 200–250 percent more precipitation than the average winter from 1965 to 2010, whereas the remaining winters in the analysis period were less than 150 percent of the long-term average.

Wells with water-level trends in group *A* are found in the northwestern part of the study area (fig. 21). Here, trends are characterized by relatively large water-level changes (fig. 20), an overall rising pattern from 1987 to 2011, and rapid responses to recharge (fig. 22). Within six months of the wet winters of 1995 and 2005, water levels in wells *WW-2 (3422 ft)* and *UE-10j (2232–2297 ft)* were rising. The overall rise for the 2005 recharge event was large (about 2 ft), but within two years, water levels were beginning to decline (fig. 20). The water-level trend suggests that the carbonate block to which the wells are open is partially isolated from the

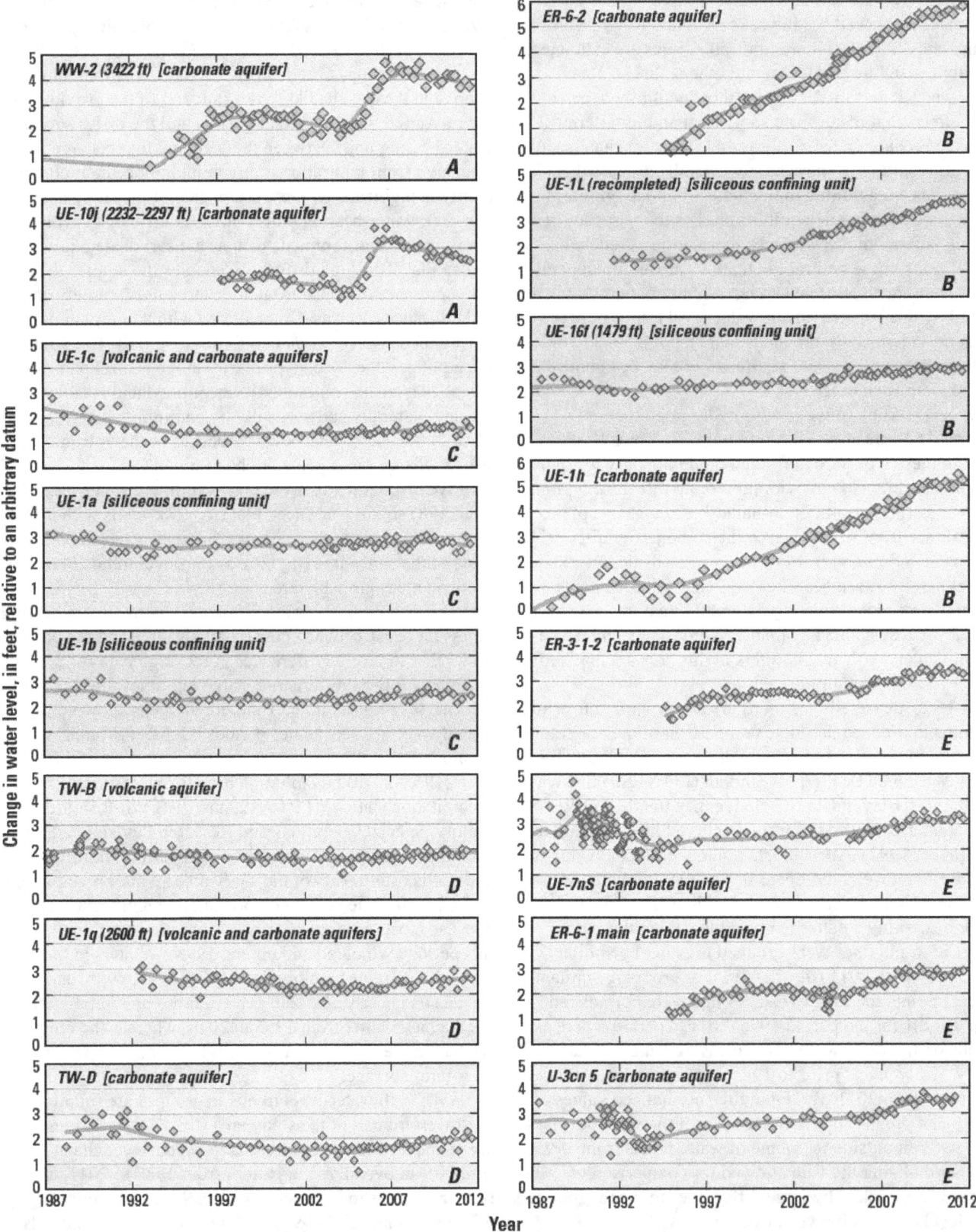

Letter in lower right is trend grouping shown on figures 21 and 22. Horizontal and vertical scales are the same on all hydrographs. Well locations are on figure 21

Figure 20. Water levels from 1987–2011 and smooth lines for wells with long-term hydrographs in Yucca Flat, Nevada. Hydrograph trends are assumed to result primarily from natural causes.

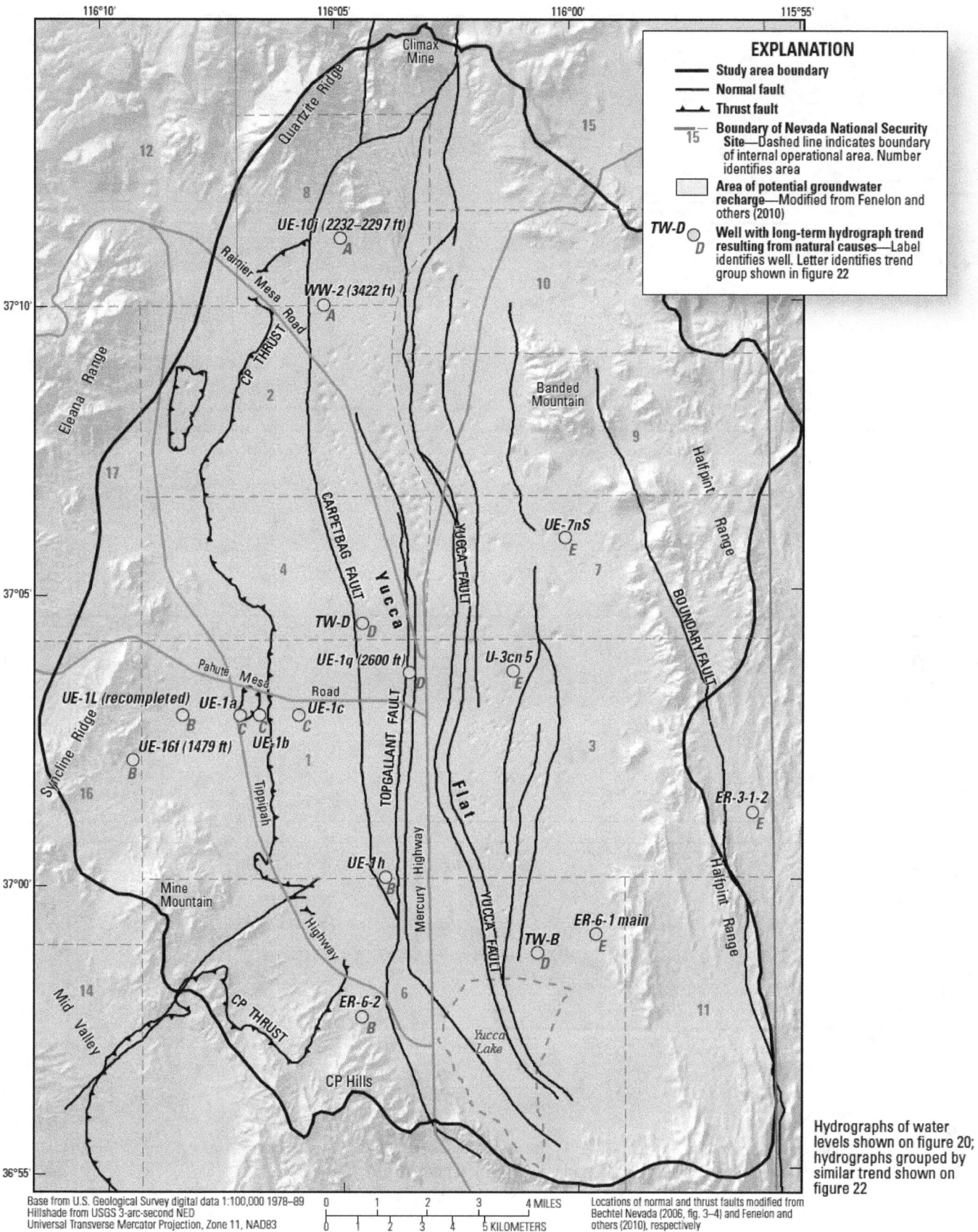

Figure 21.   Wells with long-term hydrograph trends resulting primarily from natural causes, Yucca Flat, Nevada.

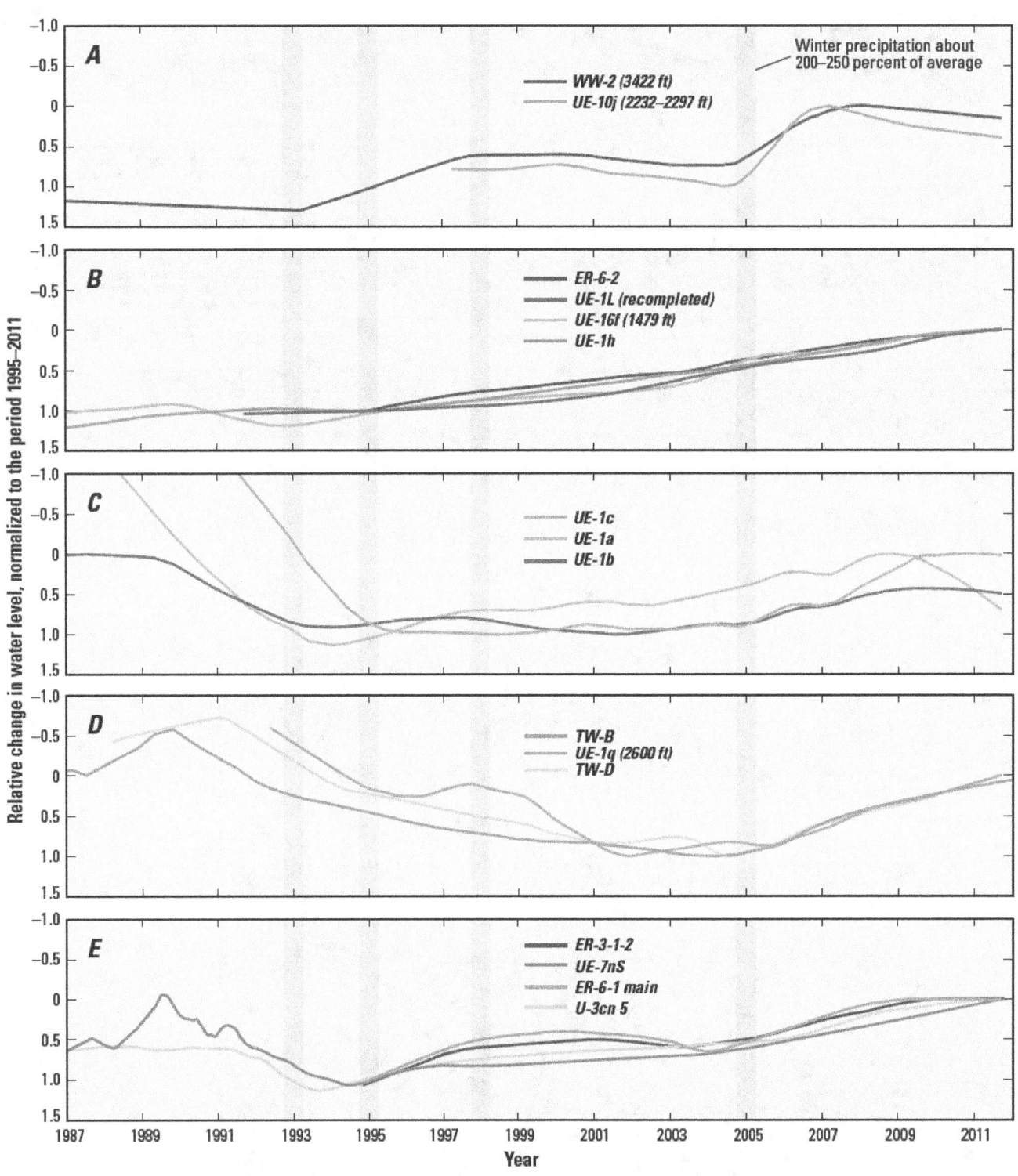

Figure 22. Smooth lines, normalized to the period 1995–2011, for water levels from wells with long-term hydrographs in Yucca Flat, 1987–2011. Hydrograph trends are assumed to result primarily from natural causes.

downgradient flow system. Isolation allows heads to quickly build up but slowly drain after a recharge event. The sharp rises suggest a good hydraulic connection to the recharge source. A possible location for recharge to enter the regional carbonate aquifer is about a mile northeast of borehole UE-10j where carbonate rock is exposed at land surface and might capture surface-water runoff from the north or northwest. Geochemical analyses of water from wells *WW-2 (3422 ft)* and *UE-10j (2232–2297 ft)* confirm that water from both wells is dominated by local recharge (Farnham and others, 2006). Water in *UE-10j (2232–2297 ft)* is composed of more than 50 percent local recharge with the remainder being a carbonate-rock geochemical signature. Similarly, *WW-2 (3422 ft)* is dominated by local recharge but has a component of ground-water with a volcanic-rock signature. The source of the volcanic-rock signature in *WW-2 (3422 ft)* likely is the overlying saturated volcanic rocks, which exhibit a steep downward hydraulic gradient (pls. 3, 4). Groundwater derived through leakage from the overlying volcanic system is expected in this well, but not in *UE-10j (2232–2297 ft)* where no overlying saturated volcanic rocks are present (pl. 4).

The water-level trend in group *B* occurs in the southwestern part of the study area (fig. 21) and is characterized by a nearly steady rising trend from 1987 to 2011 (figs. 20, 22). These trends are evident in both the siliceous confining unit and the carbonate aquifer. The water-level rises in the carbonate aquifer (6 ft in well *ER-6-2* from 1995 to 2011) are the largest naturally occurring changes documented in the study area. The rise in well *UE-16f (1479 ft)* is small (about 1 ft), but is consistent with the magnitude of changes observed in most wells open to the siliceous confining unit. Although subtle, the rate of rise increased in the mid-1990s and was maintained through 2010. The consistent rise since 1987 suggests a relatively long-term wetter-than-normal period. Although the elapsed time that occurs between a recharge event and a groundwater response in this area is unknown, the latter part of the twentieth century was wetter than the earlier part in south-central Nevada (Fenelon and Moreo, 2002, fig. 7). Mean annual precipitation from 1964 to 2011 was 8.5 in/yr in south-central Nevada, which is 23 percent greater than from 1900 to 1964 (Western Regional Climate Center, 2012). Additionally, in order to maintain a long-term rise, the wells likely are located in a semi-isolated groundwater system that drains slowly. This trend is consistent with these wells being located in the siliceous confining unit near recharge sources on Syncline Ridge and Mine Mountain (fig. 21). Well *ER-6-2* is open to a thrust block of carbonate aquifer underlain by the siliceous confining unit (section *F–F'*), whereas well *UE-1h* lies immediately east of Carpetbag fault. Water levels in both wells probably are affected by these and other structures that compartmentalize the flow system (Fenelon and others, 2010). The water chemistries in wells *UE-1h* and *ER-6-2* are similar and can be explained with a component water consisting of paleo-recharge and water from the siliceous confining unit to the west; a third component with a carbonate signature is necessary for *ER-6-2* (Farnham and others, 2006). Components of

paleo-recharge and water from the siliceous confining unit are consistent with the concept of water moving slowly through this unit from the west.

Wells with hydrograph trends in group *C* are located in the west-central part of the study area (fig. 21). Hydrographs in this group are characterized by small overall water-level changes of less than about 1 ft from 1987 to 2011 (fig. 20) and by large declining trends in the early part of the record relative to small rising trends from about 1995 to 2011 (fig. 22). The relatively large declining trends through the early 1990s represent water-level changes of only about 1 ft, but the changes appear magnified when data are normalized because the overall records contain almost no variation. The early-record trend is suspect because the magnitude of the water-level change is about the same as the measurement error. Measurement error in these wells was reduced from about 0.5–1 ft to 0.1 ft beginning in 1991 for wells *UE-1a* and *UE-1b* and in 1996 for well *UE-1c* with the use of a more accurate measurement method (Elliott and Fenelon, 2010). Therefore, the trends in group *C* may be similar to trends in group *B* or another group. Interestingly, the wells in group *C* include *UE-1c*, which is open to the volcanic and carbonate aquifers, and *UE-1a* and *UE-1b*, which are open to the siliceous confining unit. A similar trend in the aquifers and confining unit suggests that the aquifers penetrated by well *UE-1c* may not be very transmissive or they are isolated from the more transmissive regional carbonate aquifer.

Wells with hydrograph trends in group *D* are located in the south-central part of the study area (fig. 21) and are open to volcanic and carbonate aquifers (fig. 20). These wells are characterized by a small overall change in water level (fig. 20) and by a long declining trend through 2005 followed by a rising trend (fig. 22). The explanation for the declining trend is not obvious but the trend may indicate a slow draining of a system that responds only to the wettest of winters, such as 1978 (not shown) and 2005 (Air Resources Laboratory, Special Operations and Research Division, 2011). The rise in 2005 is coincident with the wet winter of 2005 and possibly is a response to recharge during this winter.

Water in wells *TW-D* and *UE-1q (2600 ft)* from group *D* are chemically similar, consisting of a large component of water with a volcanic-rock signature and a minor component of water with a carbonate-rock signature (Farnham and others, 2006). Furthermore, Farnham and others (2006) suggest that the source of the carbonate water in these wells is dissimilar to water from northern Yucca Flat. Saturated volcanic rocks above and to the west of these wells are expected to contribute water to the carbonate aquifer in this area (pl. 3). Interestingly, both wells are adjacent to major faults (pl. 4; section *CC–CC'*, pl. 2), which could provide vertical pathways for water to move down into the regional carbonate aquifer. Because hydraulic gradients suggest flow to these two wells is from the west (pl. 4), a carbonate-rock signature different from that in northern Yucca Flat is not surprising. A potential source of water with a distinct carbonate-rock signature is the isolated carbonate aquifer at Syncline Ridge. This unit appears to be laterally connected to the volcanic aquifer to the east and is upgradient of both *UE-1q (2600 ft)* and *TW-D* (section *C–C'*, pl. 1).

Hydrograph group E includes wells open to the carbonate aquifer east of Yucca fault along and east of the central axis of the basin (fig. 21). The magnitude of water-level change is moderate for these wells. Trends are characterized by an early period of decline followed by an overall rising trend with noticeable steeper rises occurring around 1995 and 2005. These rises correspond with the wet winters of 1995 and 2005 and appear similar to, but less amplified than, trends in group *A*. As discussed in the previous section, "Pumping", the trends in group E appear to correlate with pumping patterns in the carbonate aquifer (fig. 19). It is likely that these water-level trends result from a combination of changes in recharge and pumping patterns.

Similarities in water chemistry, especially strontium isotopic compositions, in wells *UE-7nS, U-3cn 5, ER-7-1, ER-6-1 main*, and *ER-6-1-2 main* suggest that these wells are along common flow paths (Farnham and others, 2006). The similarities in chemistry and hydrograph trends of these wells and the good hydraulic connection, as demonstrated in the *ER-6-1-2 main* aquifer test, strongly suggest a common southerly flow path along fault-controlled pathways. Water compositions in the aforementioned wells were explained by Farnham and others (2006) as a mixture of carbonate-rock water, volcanic-rock water, and modern recharge. Modern recharge likely enters from outcrops of carbonate rock in the Halfpint Range to the east (fig. 2) and water with a volcanic-rock signature may be contributed from proximal or overlying volcanic rocks.

Flow Conceptualization

Groundwater flow in Yucca Flat is influenced primarily by the distribution of geologic units with highly contrasting permeabilities, major faults that create barriers and conduits, and limited recharge. The interrelations between these controls determine flow directions and rates and the flow interactions between aquifer systems. This section provides a summary of the flow conceptualization for Yucca Flat. Justifications for statements made here are provided in earlier sections of the report.

Two aquifer systems are conceptualized in Yucca Flat. The alluvial–volcanic aquifer system consists of alluvial and volcanic aquifers and a volcanic confining unit (pls. 1, 3). The alluvial and volcanic aquifers occur at the water table and function as a single, typically unconfined aquifer. Throughout most of its extent, this unconfined aquifer overlies the volcanic confining unit, which provides a hydraulic barrier to the underlying regional carbonate aquifer. The majority of the alluvial–volcanic aquifer system lies in central Yucca Flat, where basin-fill deposits are thickest and most of the underground nuclear testing in Yucca Flat occurred. The carbonate aquifer system includes the regional carbonate aquifer, local carbonate

aquifers, and a siliceous confining unit (pls. 1, 4). The local carbonate aquifers consist of isolated blocks of carbonate rock that generally drain to the surrounding siliceous confining unit. The regional carbonate aquifer is present in all but the northeastern corner of the study area, where erosion has removed this aquifer and exposed siliceous rocks that form the underlying regional basement confining unit. The regional aquifer is confined in the central part of Yucca Flat where it is overlain by the volcanic confining unit, and the regional aquifer is typically unconfined elsewhere. The siliceous confining unit consists of siliciclastic and granitic rocks and forms the lateral and lower boundaries for the carbonate aquifers throughout most of Yucca Flat.

Hydraulic heads in the alluvial–volcanic aquifer are elevated about 20–100 ft relative to heads in the underlying regional carbonate aquifer. The elevated heads result from the intervening, low-permeability volcanic confining unit that isolates this shallower aquifer and restricts flow to the regional carbonate aquifer. Heads in this shallower aquifer, where water is semi-trapped by the intervening confining unit, become elevated relative to the faster draining carbonate aquifer. Similarly, hydraulic heads in the volcanic confining unit, in areas where the confining unit is exposed at the water table, commonly are elevated about 10–100 ft relative to the alluvial–volcanic aquifer. In the absence of direct infiltration of modern-day precipitation, the elevated heads in the aquifer and confining unit suggest a very slow-responding system. These heads may be elevated because the low hydraulic conductivity of the volcanic confining unit greatly delays equilibration from higher water tables during the late-Pleistocene. Furthermore, where no alluvial–volcanic aquifer is present, confining unit heads may be elevated as a result of infiltration of paleo-recharge that could be occurring currently through the thick unsaturated zone. Water in the volcanic confining unit is believed to be draining outward toward the surrounding alluvial–volcanic and carbonate aquifers.

Recharge to the alluvial–volcanic aquifer is limited, likely less than about 65 acre-ft/yr. The limited recharge is derived from slow drainage of water out of the volcanic confining unit, inflows from the carbonate aquifer in small areas where the carbonate aquifer is hydraulically connected to the alluvial–volcanic aquifer, and possibly from infiltration of paleo-recharge that still is slowly reaching the water table through the approximately 1,500-ft thick unsaturated zone. Recharge from direct infiltration of modern-day precipitation is assumed to be minor.

Groundwater discharge from the alluvial–volcanic aquifer is to the regional carbonate aquifer and is controlled by limited hydraulic connections with the carbonate aquifer. For the alluvial–volcanic aquifer in central Yucca Flat, these connections likely occur (1) near Carpetbag fault on the western side of the aquifer, (2) along parts of Yucca fault where the alluvial–volcanic aquifer is juxtaposed against the carbonate aquifer, and (3) on the northern end of the aquifer where the volcanic confining unit thins and may not provide an effective barrier to downward flow. These discharge locations allow for slow drainage of the alluvial–volcanic aquifer and largely control the lateral and vertical movement of groundwater flow within. The presence of Yucca fault results in a U-shaped flow path (south to east to north) in the southern part of the aquifer, where the fault forms a barrier along part of its extent and provides a conduit for flow into the carbonate aquifer further north. In the northern part of the aquifer, flow is to the north-northwest and discharges to the carbonate aquifer where the confining unit likely thins or is absent.

Conceptualization of flow in the regional carbonate aquifer is restricted to the shallow part where the saturated carbonate aquifer is less than about 6,000 ft below land surface. The conceptualization focuses on the shallow part because any radionuclides in the saturated zone likely will remain at relatively shallow depths as they are transported toward down-gradient discharge areas. The carbonate aquifer is bounded on the northwest, north, and northeast by a low-permeability siliceous confining unit. The western side of the aquifer is bounded by a thick wedge of siliceous confining unit that overlies a deep extension of the regional carbonate aquifer and restricts flow in the shallow part of the carbonate aquifer. The shallow regional carbonate aquifer extends outside the study area to the south, southeast, and southwest.

The total amount of groundwater flow through Yucca Flat is relatively minor and estimated to be 1,000 acre-ft/yr or less. This flow includes minor amounts of water draining into the regional carbonate aquifer from the alluvial–volcanic aquifer system. In addition, the carbonate aquifer accepts minor inflows from the northwest, north, and northeast, but the low permeability of the surrounding confining unit limits inflows from these directions. The most significant inflows likely occur from the west and southeast. In the highlands on the west side of Yucca Flat, recharge from infiltration of precipitation enters the regional carbonate aquifer. In the southeastern part of the study area, flow may enter through the part of the carbonate aquifer that extends outward to the east.

The various sources of flow to the regional carbonate aquifer converge to form a principal southward flow path through Yucca Flat centered near Yucca fault (pl. 4). Flow directions in the aquifer west of Carpetbag fault are uncertain, but they are assumed to be eastward toward the center of Yucca Flat. This western area contains a large thrusted section of carbonate aquifer that is assumed to be connected hydraulically to the regional carbonate aquifer to the east. Flow from the northeastern part of Yucca Flat is assumed to be southwestward, also toward the center of Yucca Flat. Potentiometric contours form an inverted V-shape pattern, with flow converging in a broad potentiometric trough down the central axis of Yucca Flat. The trough, as defined by the 2,390-ft contour, is portrayed as being about 3 mi wide and 15 mi long. The narrow trough is the result of preferential flow along major faults that create high-permeability pathways in a north-south direction. It is not known how important individual faults are to controlling flow paths, but it is conceptualized that the central part of Yucca Flat contains a network of highly fractured rock that interconnects the various faults in this area. The network

of parallel north-south faults likely creates a fault zone with anisotropic conditions that promote a strong preference for southerly flow through the area.

The groundwater flow system in Yucca Flat is considered to be relatively stagnant because of limited direct recharge and relatively impermeable boundaries on the west, north, and northeast that restrict groundwater flow into the basin. The small amounts of water flowing through the transmissive, fractured, regional carbonate aquifer result in a flat horizontal hydraulic gradient and relatively slow movement of water. These concepts are consistent with the long travel-time estimates, based on carbon-14 data, from Farnham and others (2006). These authors report travel times through the regional carbonate aquifer from northern to southern Yucca Flat that range from about 16,000 to 24,000 years. This equates to an average linear velocity through the carbonate potentiometric trough of about 5–6 ft/yr. Travel-time estimates are reported to be even longer for water in the alluvial–volcanic aquifer to move vertically through the volcanic confining unit and then laterally through the regional carbonate aquifer. Travel times to *WW-C (recompleted)*, when starting from wells *ER-2-1 main (shallow)* or *TW-B*, are estimated to range from about 24,000 to 35,000 years. Only in the southeastern corner of the study area are velocities reported to be higher (14–43 ft/yr) in the carbonate aquifer (Farnham and others, 2006).

Water in the regional carbonate aquifer flows toward areas of progressively lower hydraulic head at the southern end of the study area and ultimately discharges at points southwest of the study area. Exiting water likely flows into the Ash Meadows flow system and discharges at springs in Ash Meadows; some of the southerly flowing water may be diverted westward through Rock Valley where it ultimately discharges to land surface in the southern Amargosa Desert or Death Valley (Fenelon and others, 2010).

The flow conceptualization of Yucca Flat, as outlined above, is for predevelopment conditions but would be nearly identical for current conditions. From 1952 to 2010, about 18,000 acre-ft of water was pumped from the alluvial and carbonate aquifers, primarily for water supply. The average rate of pumping during this period, 310 acre-ft/yr, is about 30–90 percent of the annual estimated flow through Yucca Flat. Pumping the alluvial aquifer resulted in relatively large water-level changes that were restricted to a localized area around the pumping well. Pumping the local carbonate aquifer resulted in localized water-level declines and recoveries, whereas pumping the regional carbonate aquifer resulted in more widespread, but small, water-level declines. Overall, pumping did not alter significantly predevelopment flow directions.

Much of the pumped water was used to support activities associated with underground nuclear testing that occurred throughout the study area from 1957 to 1992. Nearly all tests were detonated in alluvial and volcanic deposits and only about 10 percent of all tests were detonated below the water table (Stoller-Navarro Joint Venture, 2009). Specific groups of tests caused large and long-lasting water-level changes as a result of pore-water pressurization in the volcanic confining unit. In the more permeable aquifers, changes to water levels following a detonation were short-lived. The overall effect of nuclear testing on the groundwater system in Yucca Flat is considered minimal in that the more substantial effects from testing are local and restricted primarily to the volcanic confining unit, which contributes minimally to groundwater flow.

Transient water-level changes resulting from natural variations in precipitation recharge occur throughout the study area but are most evident in the carbonate aquifers and are most pronounced in the northwestern and southwestern parts of the study area near bordering highland areas. Measured changes over the past 25 years do not exceed 6 ft. Responses from wet winters can be rapid and show up in the regional carbonate aquifer in less than 1 year following a wet winter. These quick responses to recharge or pumping at large distances from the stress demonstrate the high diffusivity of the regional carbonate aquifer.

Limitations and Considerations

The potentiometric surface maps, hydrogeologic sections, transient water-level analysis, and accompanying datasets in this report support a conceptual model of flow in the aquifer systems for Yucca Flat, Nevada National Security Site (NNSS). The focus is primarily on shallow groundwater flow paths, most likely to influence groundwater transport of radionuclides introduced by underground nuclear testing in Yucca Flat. The detailed flow conceptualization in this report fits within and is consistent with the conceptual framework of the regional flow system presented in Fenelon and others (2010). The results presented here serve as a basis for future work at Yucca Flat, including investigations directed at environmental restoration, underground nuclear testing, and development of water supplies. This report may be especially useful as a source of hydraulic-head data, potentiometric surface configuration, and flow-system concepts for groundwater model development and calibration.

The conceptualization of groundwater flow presented in this report is limited by geologic and hydrologic data deficiencies and simplifying assumptions regarding the geologic framework and hydrologic flow system. The geologic foundation used to delineate the extents of aquifers and confining units identified in this report is from three-dimensional hydrostratigraphic framework models (Bechtel Nevada, 2006; National Security Technologies, 2007a). The geologic framework was simplified here by grouping high- and low-permeability rocks into several major types of aquifers and confining units representing two aquifer systems. This simplification portrays the aquifer systems as composed of two distinct aquifers bounded and separated by confining units, when in reality the mix and diversity of geologic materials represents a continuum that ranges from highly transmissive to nearly impermeable units.

The analysis is focused primarily on the shallow flow system, defined as extending to a depth of 6,000 ft below land surface and where all well data were collected. The deep part of the hydrologic system, assumed to be less active with flow rates much lower than those in the shallow system, exerts minimal influence on the transport of radionuclides off the NNSS. However, very young faults and seismically active and potentially active faults, such as the north-striking normal faults in Yucca Flat, may enhance permeability at depth (Potter and others, 2002). These fault-enhanced pathways could allow for shallow-to-deep or deep-to-shallow flow along fault conduits. Potential for deep-system flow interactions with the shallow system along seismically active faults was not addressed as part of this study.

In developing the flow conceptualization presented in this study, assumptions were made about the significance of structures controlling groundwater flow. In some cases, data support the interpretation that a structure, such as Yucca fault, controls flow; in other cases, similar interpretations are based only on assumptions about the extent, permeability, or ability of the structure to juxtapose geologic units. The north-striking normal faults in Yucca Flat create a transmissive corridor within the carbonate aquifer, but there have been few attempts to identify specific active faults, determine their individual flow properties, or define their relationship to the potentiometric surface. The hydraulic trough in the center of Yucca Flat is portrayed as a several-mile wide transmissive corridor of highly fractured rock and interconnected faults rather than as a more discrete feature controlled by only a few highly transmissive fault planes, even though this is a possibility. CP thrust and Carpetbag fault are examples of potentially significant structures that could act as flow barriers; however, few data are available to support this assumption. Fenelon and others (2010) portrayed groundwater in the regional carbonate aquifer bounded by these two faults as flowing southward, with Carpetbag fault providing a barrier to eastward flow. This report portrays an increased hydraulic gradient in the regional carbonate aquifer west of Carpetbag fault to suggest a slight impedance to flow across the fault. However, the dominant flow direction portrayed west of Carpetbag fault is eastward across the fault and into the center of Yucca Flat. Eastward or southward (Fenelon and others, 2010) flow in this thrusted rock is plausible, given the limited data available. The direction of flow has potential implications for radionuclide transport because of the many underground nuclear tests conducted in the western part of Area 2 of the NNSS.

Either of the alternative 2,380-ft contours for the regional carbonate aquifer (pl. 4) deviate from the conceptualization of Fenelon and others (2010). The 2,380-ft contour from Fenelon and others (2010) was broader and centered so that the potentiometric low in the trough was between boreholes ER-3-1 and ER-6-1. This conceptualization shows the low further west, between ER-6-1 and WW-C or just west of WW-C, depending on the alternative. The primary reason for shifting the 2,380-ft contour westward is because of a new hydrograph interpretation (Elliott and Fenelon, 2010) for well *ER-5-3-2*, located

10 mi south of borehole ER-3-1. Water levels in *ER-5-3-2* rose for 10 years following well completion, development, and testing. This rise was re-interpreted to reflect equilibration from these activities rather than a natural rise due to climatic conditions. As a result, the hydraulic-head estimate in well *ER-5-3-2* is 9 ft higher than reported in Fenelon and others (2010) and necessitates shifting the 2,380-ft contour line to the west. The westward shift results in a narrower trough centered closer to Yucca fault. The exact center of the trough is unknown, as exemplified by the two alternative contours. One or two wells drilled into the carbonate aquifer to the west or east of WW-C would provide insight on the horizontal hydraulic gradient and the location of the potentiometric low. If the conceptualization of a narrow trough is correct, then all water moving out of Yucca Flat may have to travel through a small corridor and a short line of wells may be all that is needed to monitor the water moving south out of Yucca Flat.

Particular areas of uncertainty related to the thrusted carbonate aquifer west of Carpetbag fault, as delineated in plate 4, include (1) the conceptualized isolation of the local carbonate aquifer centered around borehole UE-2ce, (2) the assumed continuity of the carbonate rock within the thrusted section between CP thrust and Carpetbag fault, (3) the inclusion of the thrusted carbonate aquifer between CP thrust and Carpetbag fault as part of the regional carbonate aquifer rather than an isolated local aquifer, and (4) the inferred hydraulic connection of the thrusted regional carbonate aquifer with the regional carbonate aquifer east of Carpetbag fault. Refinement of these hydrologic concepts will require additional subsurface geologic and hydrologic data.

Within the alluvial–volcanic aquifer, flow directions are uncertain in several areas listed below.

- Portrayal of eastward flow in the western alluvial–volcanic aquifer is highly uncertain because data are limited to a single well, *UE-1c*, which is open to a composite of volcanic and carbonate aquifers.
- The southern alluvial–volcanic aquifer has adequate data in most areas to support flow interpretations, but the portrayal of flow in the northwestern part of this aquifer (between wells *U-4au* and *UE-1k*, pl. 3) is uncertain. Contours in this area suggest high gradients and southerly and westerly flow but an alternative interpretation is that the aquifer in this area is somewhat compartmentalized.
- The interpretation of Yucca fault as a flow barrier in the southern alluvial–volcanic aquifer is supported by hydraulic-head data in the area of boreholes UE-1k and ER-3-2 (pl. 3). Further south, hydraulic-head data in the area of WW-3 and UE-6e suggest the fault is not a barrier. The location of the transition from a barrier to no-barrier is uncertain.
- Flow directions portrayed in the northern alluvial–volcanic aquifer are supported by most of the head data, but sufficient anomalous heads exist to make flow interpretation uncertain. Additionally, a drain to the carbonate aquifer on the northern end of the alluvial–volcanic aquifer is assumed to exist but the location is unknown.

Hydraulic connections are portrayed along Carpetbag fault (pl. 2) that allow lateral flow between the alluvial–volcanic and carbonate aquifers. These hydraulic connections are inferred from aquifer-on-aquifer connections in the Yucca Flat hydrostratigraphic framework model (Bechtel Nevada, 2006). Where modeled connections exist, directions of flow are inferred from assumed hydraulic-head distributions in the two aquifers (pls. 3, 4). The southern alluvial–volcanic aquifer is interpreted to receive some flow from the carbonate aquifer along Carpetbag fault in the southern part of the aquifer and to lose some flow to the carbonate aquifer in the northern part (pl. 3). Further north, the northern alluvial–volcanic aquifer is interpreted to lose flow to the carbonate aquifer along the Carpetbag fault. These interpretations are highly uncertain, primarily because of uncertainties in the head distribution of the thrusted carbonate aquifer west of the fault. In the southern part of aquifer system, heads in the two aquifers likely are similar and small amounts of flow could be moving in either direction depending on the exact head distributions and location. Further north, if the alluvial–volcanic aquifer is losing water to the carbonate aquifer, as portrayed, the amount of flow likely is limited. It is possible that along the Carpetbag fault, especially the northern part, the area of hydraulic connection is much more limited than portrayed in the framework models. If so, then flow interactions between the two aquifer systems also would be minimal.

Potentiometric contours and flow arrows shown on the plates are intended to portray general directions of groundwater flow. The effects of anisotropy on mapped flow directions were accounted for indirectly in this analysis. At the local scale, however, anisotropy can cause flow to take tortuous paths that may differ from the regional flow direction. At a more regional scale, faults and fracture zones form flow barriers or preferred pathways, creating large-scale anisotropy that also can result in flow paths that deviate from the directions implied by the potentiometric contours. It is possible that flow is concentrated along a series of north-south faults and fractures that are semi-isolated from each other. If true, the smooth contours that suggest flow toward the center of the trough would be, in fact, a series of step-like drops in head that reflect the compartmentalization of the regional carbonate aquifer into parallel, north-south mini-flow systems.

Vertical hydraulic gradients are accounted for by portraying independent sets of heads for the alluvial–volcanic and carbonate aquifers. The assumption was made that within a mapped aquifer, vertical gradients were negligible relative to horizontal gradients. However, the volcanic confining unit is different from the aquifers and most flow in this unit is thought to be vertical. The intent of the contours in the confining unit is to portray the elevated hydraulic heads in this unit that clearly demonstrate that the confining unit is nearly impermeable, drains into and is a source of water for the alluvial–volcanic aquifer, and effectively compartmentalizes flow within the alluvial–volcanic aquifer.

Other limitations include a lack of hydraulic-head data in parts of the study area and uncertainty associated with the assumptions used to differentiate between predevelopment and nuclear-test affected heads. For the alluvial–volcanic aquifer, hydraulic-head data are sparse in the western part of the study area. For the carbonate aquifer, head data are sparse west of Carpetbag fault and in the northeastern part of the study area. Lack of head data created particular problems in areas where hydraulic continuity between parts of aquifers was believed to be impeded, such as along Carpetbag fault. In these areas, prediction of the head in a potentially isolated aquifer was difficult because the degree of isolation of the aquifer was unknown and, therefore, nearby head data could not necessarily be used to predict the head. Where water-level measurements were available, their conversion to a predevelopment hydraulic head representing a specific aquifer was based on assumptions that the well is open only to the hydrologic unit that was targeted by the completion and that the water level represents natural conditions in the targeted hydrologic unit. This second assumption was difficult to verify where water-level measurements in a well did not show a clear trend with time. Where water-level data were limited or wells were open to confining units that equilibrate slowly, the potential for misidentifying a nonstatic or nuclear-test affected head for a predevelopment head was high. Discerning a predevelopment head from a nuclear-test affected head was especially difficult in the volcanic confining unit because a sustained elevated head can result from natural conditions or from nuclear testing.

Funding for this study was provided by the U.S. Department of Energy Office of Environmental Management, National Nuclear Security Administration, Nevada Site Office, under Interagency Agreement DE-NA0001654. Robert McFaul and Mike Pinto of Dynamic Graphics, Inc., developed numerical scripts and initial datasets and provided technical guidance in the construction of three-dimensional cavity-radius spheres using the Dynamic Graphics EarthVision™ geologic modeling software. Vertical sections through the computed spheres are depicted on plate 2 of this report.

Keith Halford (U.S. Geological Survey) developed the Microsoft® Excel macro programming within the interactive Excel Workbook in appendix 3. Keith's analyses of aquifer-test results in Yucca Flat and development of simple numerical models helped to guide our interpretation of flow in the alluvial–volcanic and regional carbonate aquifer systems.

References Cited

Air Resources Laboratory, Special Operations and Research Division, 2011, Nevada Test Site (NTS) climatological rain gauge network: National Oceanic and Atmospheric Administration, accessed December 2011 at *http://www.sord. nv.doe.gov/home_climate_rain.htm*.

Asch, T.H., Sweetkind, D., Burton, B.L., and Wallin, E.L., 2009, Detailed geophysical fault characterization in Yucca Flat, Nevada Test Site, Nevada: U.S. Geological Survey Open-File Report 2008-1346, 64 p.

Bechtel Nevada, 2004, Completion report for well ER-8-1: U.S. Department of Energy Report DOE/NV/11718–854, 105 p.

Bechtel Nevada, 2006, A hydrostratigraphic model and alternatives for the groundwater flow and contaminant transport model of Corrective Action Unit 97—Yucca Flat–Climax Mine, Lincoln and Nye Counties, Nevada: U.S. Department of Energy Report DOE/NV/11718–1119, 288 p.

Black, J.H., Alexander, J., Jackson, P.D., Kimbell, G.S., and Lake, R.D., 1987, The role of faults in the hydrogeological environment: Report of Fluid Processes Research Group, British Geological Survey, 54 p.

Blainey, J.B., Webb, R.H., and Magirl, C.S., 2007, Modeling the spatial and temporal variation of monthly and seasonal precipitation on the Nevada Test Site and vicinity, 1960–2006: U.S. Geological Survey Open-File Report 2007–1269, 40 p.

Blankennagel, R.K., and Weir, J.E., Jr., 1973, Geohydrology of the eastern part of Pahute Mesa, Nevada Test Site, Nye County, Nevada: U.S. Geological Survey Professional Paper 712–B, 35 p.

Bowen, S.M., Finnegan, D.L., Thompson, J.L., Miller, C.M., Baca, P.L., Olivas, L.F., Geoffrion, C.G., Smith, D.K., Goishi, Wataru, Esser, B.K., Meadows, J.W., Namboodiri, Neil, and Wild, J.F., 2001, Nevada Test Site radionuclide inventory, 1951–1992: Los Alamos National Laboratory Report LA–13859–MS, 28 p.

Burkhard, N.R., and Rambo, J.T., 1991, One plausible explanation for water mounding: Proceedings of the Sixth Symposium on Containment of Underground Nuclear Explosions, Reno, Nevada, University of Nevada, Sept. 24–27, 1991, CONF–9109114, v. 2, p. 145–158.

Byers, F.M., Jr., Carr, W.J., Orkild, P.P., Quinlivan, W.D., and Sargent, K.A., 1976, Volcanic suites and related cauldrons of the Timber Mountain-Oasis Valley caldera complex, southern Nevada: U.S. Geological Survey Professional Paper 919, 70 p.

Caine, J.S., Evans, J.P., and Forster, C.B., 1996, Fault zone architecture and permeability structure: Geology, v. 24, p. 1,025–1,028.

Caine, J.S., and Forster, C.B., 1999, Fault zone architecture and fluid flow—Insights from field data and numerical modeling, in Haneberg, W.C., Mozley, P.S., Moore, J.C., and Goodwin, L.B., eds., Faults and subsurface fluid flow in the shallow crust: American Geophysical Union Monograph 113, p. 101–127.

Cardinalli, J.L., Roach, L.M., Rush, F.E., and Vasey, B.J., 1968, State of Nevada hydrographic areas: Nevada Division of Water Resources map, scale 1:500,000.

Carle, S.F., Zavarin, Mavrik, Sun, Yunwei, and Pawloski, G.A., 2008, Evaluation of hydrologic source term processes for underground nuclear tests in Yucca Flat, Nevada Test Site—carbonate tests: Lawrence Livermore National Laboratory Report LLNL–TR–403485, 372 p.

Carr, W.J., Byers, F.J., Jr., and Orkild, P.P., 1986, Stratigraphic and volcano-tectonic relations of Crater Flat Tuff and some older volcanic units, Nye County, Nevada: U.S. Geological Survey Professional Paper 1323, 28 p.

Caskey, S.J., and Schweickert, R.A., 1992, Mesozoic deformation in the Nevada Test Site and vicinity: Framework of the Cordilleran fold and thrust belt and Tertiary extension north of Las Vegas Valley: Tectonics, v. 11, p. 1,314–1,331.

Claassen, H.C., 1973, Water quality and physical characteristics of Nevada Test Site water-supply wells: U.S. Geological Survey Open-File Report 474–158, 145 p.

Cole, J.C., 1997, Major structural controls on the distribution of Pre-Tertiary rocks, Nevada Test Site vicinity, southern Nevada: U.S. Geological Survey Open-File Report 97–533, 19 p.

Cole, J.C., and Cashman, P.H., 1999, Structural relationships of pre-Tertiary rocks in the Nevada Test Site region, southern Nevada: U.S. Geological Survey Professional Paper 1607, 39 p.

Cole, J.C., Harris, A.G., and Wahl, R.R., 1997, Subcrop geologic map of pre-Tertiary rocks in the Yucca Flat and northern Frenchman Flat areas, Nevada Test Site, southern Nevada: U.S. Geological Survey Open-File Report 97–678, scale 1:48,000, 24 p.

D'Agnese, F.A., Faunt, C.C., and Turner, A.K., 1998, An estimated potentiometric surface of the Death Valley region, Nevada and California, developed using Geographic Information System and automated interpolation techniques: U.S. Geological Survey Water-Resources Investigations Report 97–4052, 15 p.

D'Agnese, F.A., O'Brien, G.M., Faunt, C.C., and San Juan, C.A., 1999, Simulated effects of the climate change on the Death Valley regional ground-water flow system, Nevada and California: U.S. Geological Survey Water-Resources Investigations Report 98–4041, 40 p.

Dettinger, M.D., Harrill, J.R., Schmidt, D.L., and Hess, J.W., 1995, Distribution of carbonate-rock aquifers and the potential for their development, southern Nevada and parts of Arizona, California, and Utah: U.S. Geological Survey Water-Resources Investigations Report 91–4146, 100 p.

Doty, G.C., and Rush, F.E., 1985, Inflow to a crack in playa deposits of Yucca Lake, Nevada Test Site, Nye County, Nevada: U.S. Geological Survey Water-Resources Investigations Report 84–4296, 24 p.

Doty, G.C., and Thordarson, William, 1983, Water table in rocks of Cenozoic and Paleozoic age, 1980, Yucca Flat, Nevada Test Site, Nevada: U.S. Geological Survey Water-Resources Investigations Report 83–4067, 1 sheet, scale 1:48,000.

Drellack, S.L., Jr., Prothro, L.B., Gonzales, J.L., and Mercadante, J.M., 2010, The hydrogeologic character of the Lower Tuff Confining Unit and the Oak Springs Butte Confining Unit in the Tuff Pile area of central Yucca Flat: U.S. Department of Energy Report DOE/NV/25946–544, 107 p.

Dudley, W.W., and Larson, J.D., 1976, Effect of irrigation pumping on desert pupfish habitats in Ash Meadows, Nye County, Nevada: U.S. Geological Survey Professional Paper 927, 52 p.

Ekren, E.B., Anderson, R.E., Rogers, C.L., and Noble, D.C., 1971, Geology of northern Nellis Air Force Base Bombing and Gunnery Range, Nye County, Nevada: U.S. Geological Survey Professional Paper 651, 91 p.

Elliott, P.E., and Fenelon, J.M., 2010, Database of groundwater levels and hydrograph descriptions for the Nevada Test Site area, Nye County, Nevada: U.S. Geological Survey Data Series 533, version 2.0, December 2011, 16 p.

Elliott, P.E., and Moreo, M.T., 2011, Groundwater withdrawals and associated well descriptions for the Nevada National Security Site, Nye County, Nevada, 1951–2008: U.S. Geological Survey Data Series 567, 126 p.

Farnham, I.M., Rose, T.P., Kwicklis, E.M., Hershey, R.L., Paces, J.B., and Fryer, W.M., 2006, Geochemical and isotopic evaluation of groundwater movement in Corrective Action Unit 97—Yucca Flat/Climax Mine, Nevada Test Site, Nevada: Stoller-Navarro Joint Venture Report S-N/99205–070, 283 p.

Faunt, C.C., 1997, Effect of faulting on ground-water movement in the Death Valley region, Nevada and California: U.S. Geological Survey Water-Resources Investigations Report 95–4132, 42 p., 1 pl.

Faunt, C.C., Sweetkind, D.S., and Belcher, W.R., 2004, Three-dimensional hydrogeologic framework model, *in* Belcher, W.R., ed., Chapter E of Death Valley regional ground-water flow system, Nevada and California—Hydrogeologic framework and transient ground-water flow model: U.S. Geological Survey Scientific Investigations Report 2004–5205, p. 165–256.

Fenelon, J.M., 2005, Analysis of ground-water levels and associated trends in Yucca Flat, Nevada Test Site, Nye County, Nevada, 1951–2003: U.S. Geological Survey Scientific Investigations Report 2005–5175, 87 p. Available at *http://pubs.usgs.gov/sir/2005/5175.*

Fenelon, J.M., Laczniak, R.J., and Halford, K.J., 2008, Predevelopment water-level contours for aquifers in the Rainier Mesa and Shoshone Mountain area of the Nevada Test Site, Nye County, Nevada: U.S. Geological Survey Scientific Investigations Report 2008–5044, 38 p.

Fenelon, J.M., and Moreo, M.T., 2002, Trend analysis of ground-water levels and spring discharge in the Yucca Mountain region, Nevada and California, 1960–2003: U.S. Geological Survey Water-Resources Investigations Report 02–4178, 97 p.

Fenelon, J.M., Sweetkind, D.S., and Laczniak, R.J., 2010, Groundwater flow systems at the Nevada Test Site, Nevada: A synthesis of potentiometric contours, hydrostratigraphy, and geologic structures: U.S. Geological Survey Professional Paper 1771, 54 p., 6 pl.,

Fenske, P.R., and Carnahan, C.L., 1975, Water table and related maps for Nevada Test Site and Central Nevada Test Area: U.S. Energy Research and Development Administration, Nevada Operations Office Report NVO–1253–9, 18 p.

Ferrill, D.A., Stamatakos, J.A., and Sims, D., 1999, Normal fault corrugation — Implications for growth and seismicity of active normal faults: Journal of Structural Geology, v. 21, p. 1,027–1,038.

Flint, A.L., Flint, L.E., Hevesi, J.A., and Blainey, J.M., 2004, Fundamental concepts of recharge in the Desert Southwest—a regional modeling perspective, *in* Hogan, J.F., Phillips, F.M., and Scanlon, B.R., eds., Groundwater recharge in a desert environment—The Southwestern United States: American Geophysical Union Water Science and Applications Series, v. 9, p. 159-184.

Freeze, R.A., and Witherspoon, P.A., 1967, Theoretical analysis of regional groundwater flow–2. Effect of water-table configuration and subsurface permeability variation: Water Resources Research, v. 3, no. 2, p. 623–634.

Grasso, D.N., 2000, Geologic surface effects of underground nuclear testing, Yucca Flat, Nevada Test Site, Nevada: U.S. Geological Survey Open-File Report 00–176, 20 p.

Grasso, D.N., 2001, GIS surface effects archive of underground nuclear detonations conducted at Yucca Flat and Pahute Mesa, Nevada Test Site, Nevada: U.S. Geological Survey Open-File Report 01–272.

Hale, G.S., Trudeau, D.T., and Savard, C.S., 1995, Water-level data from wells and test holes through 1991 and potentiometric contours as of 1991 for Yucca Flat, Nevada Test Site, Nye County, Nevada: U.S. Geological Survey Water-Resources Investigation Report 95–4177, 1 sheet, scale 1:48,000.

Halford, K.J., Laczniak, R.J., and Galloway, D.L., 2005, Hydraulic characterization of overpressured tuffs in central Yucca Flat, Nevada Test Site, Nye County, Nevada: U.S. Geological Survey Scientific Investigations Report 2005–5211, 36 p.

Hansen, W.R., Lemke, R.W., Cattermole, J.M., and Gibbons, A.B., 1963, Stratigraphy and structure of the Rainier and USGS tunnel areas: U.S. Geological Survey Professional Paper 382–A, 49 p., 6 pl.

Harrill, J.R., and Bedinger, M.S, 2004, Estimated model boundary flows, *in* Belcher, W.R., ed., Appendix 2 of Death Valley regional ground-water flow system, Nevada and California—Hydrogeologic framework and transient ground-water flow model: U.S. Geological Survey Scientific Investigations Report 2004–5205, p. 375-408.

Harrill, J.R., Gates, J.S., and Thomas, J.M., 1988, Major ground-water flow systems in the Great Basin region of Nevada, Utah, and adjacent States: U.S. Geological Survey Hydrologic Investigations Atlas HA-694-C, 2 sheets.

Harrill, J.R., and Prudic, D.E., 1998, Aquifer systems in the Great Basin region of Nevada, Utah, and adjacent States—A summary report: U.S. Geological Survey Professional Paper 1409–A, 66 p.

Helsel, D.R., and Hirsch, R.M., 1992, Studies in environmental science 49—Statistical methods in water resources: Amsterdam, Elsevier, 522 p.

Hevesi, J.A., Flint, A.L., and Flint, L.E., 2003, Simulation of net infiltration and potential recharge using a distributed-parameter watershed model of the Death Valley region, Nevada and California: U.S. Geological Survey Water-Resources Investigations Report 03–4090, 161 p.

Hokett, S.L., Gillespie, D.R., Wilson, G.V., and French, R.H., 2000, Evaluation of recharge potential at subsidence crater U10i, northern Yucca Flat, Nevada Test Site: Desert Research Institute Publication No. 45174, 40 p.

Houser, F.N., and F.G. Poole, 1960, Preliminary geologic map of the Climax stock and vicinity, Nye County, Nevada: U.S. Geological Survey Map I-328, scale 1:4,800.

Hudson, M.R., 1992, Paleomagnetic data bearing on the origin of arcuate structures in the French Peak-Massachusetts Mountain area of southern Nevada: Geological Society of America Bulletin, v. 104, p. 581–594.

IT Corporation, 1996, Underground Test Area subproject, Phase I data analysis task, volume IV, Hydrologic parameter data documentation package: Report ITLV/10972–181 prepared for the U.S. Department of Energy, 8 v. [variously paged].

Kim, Y.-S., Peacock, D.C.P., and Sanderson, D.J., 2004, Fault damage zones: Journal of Structural Geology, v. 26, p. 503–517.

Knox, J.B., Rawson, D.E., and Korver, J.A., 1965, Analysis of a groundwater anomaly created by an underground nuclear explosion: Geophysical Research, v. 70, no. 4, p. 823–835.

Kwicklis, E.M., Wolfsberg, A.V., Stauffer, P.H., Walvoord, M.A., and Sully, M.J, 2006, Multiphase, multicomponent parameter estimation for liquid and vapor fluxes in deep arid systems using hydrologic data and natural environmental tracers: Vadose Zone Journal, v. 5, no. 3, p. 934–950.

Laczniak, R.J., Cole, J.C., Sawyer, D.A., and Trudeau, D.A., 1996, Summary of hydrogeologic controls on ground-water flow at the Nevada Test Site, Nye County, Nevada: U.S. Geological Survey Water-Resources Investigations Report 96–4109, 59 p.

Laczniak, R.J., Smith, J.L., Elliott, P.E., DeMeo, G.A., Chatigny, M.A., and Roemer, G.J., 2001, Ground-water discharge determined from estimates of evapotranspiration, Death Valley regional flow system, Nevada and California: U.S. Geological Survey Water-Resources Investigations Report 01–4195, 51 p.

Levitt, D.G., and Yucel, V., 2002, Potential groundwater recharge and the effects of soil heterogeneity on flow at two radioactive waste management sites at the Nevada Test Site: U.S. Department of Energy Report DOE/NV/11718—609, 11 p.

Levy, S.S., 1991, Mineralogic alteration history and paleohydrology at Yucca Mountain, Nevada: High-Level Radioactive Waste Management Proceedings of the Second Annual International High Level Radioactive Waste Management Conference, Las Vegas, Nev., 1991: Proceedings, American Society of Civil Engineers, p. 477–485.

Marshall, B.D., Peterman, Z.E., and Stuckless, J.S., 1993, Strontium isotopic evidence for a higher water table at Yucca Mountain: High-Level Radioactive Waste Management Proceedings of the Fourth International Conference, American Society of Civil Engineers, v. 2, p. 1,948–1,952.

McNab, W.W., 2008, Evaluation of hydrologic source term processes for underground nuclear tests in Yucca Flat, Nevada Test Site—Unsaturated tests and the impact of recharge: Lawrence Livermore National Laboratory Report LLNL-TR-403360, 95 p.

Meinzer, O.E., 1923, Outline of ground-water hydrology, with definitions: U.S. Geological Survey Water-Supply Paper 494, 71 p.

National Security Technologies, LLC, 2007a, A hydrostratigraphic model and alternatives for the groundwater flow and contaminant transport model of Corrective Action Unit 99—Rainier Mesa–Shoshone Mountain, Nye County, Nevada: U.S. Department of Energy Report DOE/NV/25946–146, 302 p.

National Security Technologies, LLC, 2007b, Characterization report, Area 3 radioactive waste management site, Nevada Test Site, Nevada: U.S. Department of Energy Report DOE/NV/25946—080, 178 p.

National Security Technologies, LLC, 2008, Completion report for the well ER-6-2 site, Correction Action Unit 97—Yucca Flat - Climax Mine: U.S. Department of Energy Report DOE/NV—1270, 103 p.

Pawloski, G.A.; Tompson, A.F.B.; Carle, S.F.; Detwiler, R.L.; Hu, Qinhong; Kollet, Stefan; Maxwell, R.M.; McNab, W.W.; Roberts, S.K.; Shumaker, D.E.; Sun, Yunwei; and Zavarin, Mavrik, 2008, Evaluation of hydrologic source term processes for underground nuclear tests in Yucca Flat, Nevada Test Site—Introduction and executive summary: Lawrence Livermore National Laboratory Report LLNL–TR–403428, 84 p.

Phelps, G.A., Jachens, R.C., Moring, B.C., and Roberts, C.W., 2004, Modeling of the Climax stock and related plutons based on the inversion of magnetic data, southwest Nevada: U.S. Geological Survey Open-File Report 2004–1345, 21 p.

Phelps, G.A., Langenheim, V.E., and Jachens, R.C., 1999, Thickness of Cenozoic deposits of Yucca Flat inferred from gravity data, Nevada Test Site, Nevada: U.S. Geological Survey Open-File Report 99–310, 33 p.

Poole, F.G., and Sandberg, C.A., 1977, Mississippian paleogeography and tectonics of the western United States, *in* Stewart, J.H., Stevens, C.H., and Fritsche, A.E., eds., Paleozoic paleogeography of the western United States, Pacific Coast Paleogeography Symposium 1: Society of Economic Paleontologists and Mineralogists, p. 67–85.

Poole, F.G., Stewart, J.H., Palmer, A.R., Sandberg, C.A., Madrid, R.J., Ross, R.J., Jr., Hintze, L.F., Miller, M.M., and Wrucke, C.T., 1992, Latest Precambrian to latest Devonian time; development of a continental margin, *in* Burchfiel, B.C., Lipman, P.W., and Zoback, M.L., eds., The Cordilleran orogen, conterminous U.S.: Geological Society of America, The Geology of North America, v. G–3, p. 9–56.

Potter, C.J., Sweetkind, D.S., Dickerson, R.P., and Killgore, M.L., 2002, Hydrostructural map of the Death Valley ground-water basin, Nevada and California: U.S. Geological Survey Miscellaneous Field Studies Map MF-2372 scale 1:350,000, 2 pl. with pamphlet.

Prothro, L.B., and Drellack, S.L., Jr., 1997, Nature and extent of lava-flow aquifers beneath Pahute Mesa, Nevada Test Site: Bechtel Nevada Report DOE/NV/11718–156 prepared for the U.S. Department of Energy, National Nuclear Security Administration Nevada Site Office, 50 p.

Prothro, L.B., Drellack, S.L., Jr., Haugstad, D.N., Huckins-Gang, H.E., and Townsend, M.J., 2009, Observations on faults and associated permeability structures in hydrogeologic units at the Nevada Test Site: National Security Technologies, LLC Report DOE/NV/25946–690 prepared for the U.S. Department of Energy, National Nuclear Security Administration Nevada Site Office, 108 p.

Quade, J., Mifflin, M.D., Pratt, W.L., McCoy, W., and Burckle, L., 1995, Fossil spring deposits in the southern Great Basin and their implications for changes in water-table levels near Yucca Mountain, Nevada, during Quaternary time: Geological Society of America Bulletin, v. 107, p. 213–230.

Rush, F.E., 1968, Index of hydrographic areas in Nevada: Nevada Division of Water Resources Information Report 6, 38 p.

Sawyer, D.A., Fleck, R.J., Lanphere, M.A., Warren, R.G., Broxton, D.E., and Hudson, M.R., 1994, Episodic caldera volcanism in the Miocene southwestern Nevada volcanic field—Revised stratigraphic framework, $^{40}Ar/^{39}Ar$ geochronology, and implications for magmatism and extension: Geological Society of America Bulletin, v. 106, no. 10, p. 1,304–1,318.

Slate, J.L, Berry, M.E., Rowley, P.D., Fridrich, C.J., Morgan, K.S., Workman, J.B., Young, O.D., Dixon, G.L., Williams, V.S., McKee, E.H., Ponce, D.A., Hildenbrand, T.G., Swadley, W.C., Lundstrom, S.C., Ekren, E.B., Warren, R.G., Cole, J.C., Fleck, R.J., Lanphere, M.A., Sawyer, D.A., Minor, S.A., Grunwald, D.J., Laczniak, R.J., Menges, C.M., Yount, J.C., and Jayko, A.S., 1999, Digital geologic map of the Nevada Test Site and vicinity, Nye, Lincoln, and Clark Counties, Nevada, and Inyo County, California: U.S. Geological Survey Open-File Report 99–554A, 53 p., 1 pl., scale 1:120,000, accessed June 2011 at *http://pubs.usgs.gov/of/1999/ofr-99-0554/*.

Snow, J.K., and Wernicke, B.P., 2000, Cenozoic tectonism in the central Basin and Range—Magnitude, rate, and distribution of upper crustal strain: American Journal of Science, v. 300, p. 659–719.

Soulé, D.A., 2006, Climatology of the Nevada Test Site: National Oceanic and Atmospheric Administration Air Resources Laboratory Technical Memorandum SORD 2006–3, 165 p.

State of Nevada, U.S. Department of Energy, and U.S. Department of Defense, 1996 (as amended May 2011), Federal Facility Agreement and Consent Order (FFACO): 855 p., accessed December 2011 at *http://ndep.nv.gov/boff/ffco.htm*.

Stewart, J. H., 1970, Upper Precambrian and Lower Cambrian strata in the southern Great Basin, California and Nevada: U.S. Geological Survey Professional Paper 620, 206 p.

Stewart, J. H., 1972, Initial deposits in the Cordilleran geosyncline; evidence of a late Precambrian (~850 m.y.) continental separation: Geological Society of America Bulletin, v. 83, p. 1,345–1,360.

Stewart, J.H., and Poole, F.G., 1974, Lower Paleozoic and uppermost Precambrian Cordilleran miogeocline, Great Basin, western United States, *in* Dickinson, W.R., ed., Tectonics and Sedimentation: Tulsa, Society of Economic Petrologists and Mineralogists, p. 27–57.

Stock, J.M., Healy, J.H., Hickman, S.H., and Zoback, M.D., 1985, Hydraulic fracturing stress measurements at Yucca Mountain, Nevada, and relationship to the regional stress field: Journal of Geophysical Research, v. 90, p. 8,691–8,706.

Stoller-Navarro Joint Venture, 2005, Analysis of hydraulic responses from the ER-6-1 multiple-well aquifer test, Yucca Flat FY 2004 testing program, Nevada Test Site, Nye County, Nevada: Stoller-Navarro Joint Venture Report S–N/99205–051, 84 p.

Stoller-Navarro Joint Venture, 2006, Phase I Hydrologic data for the groundwater flow and contaminant transport model of Corrective Action Unit 97: Yucca Flat/Climax Mine, Nevada Test Site, Nye County, Nevada: Stoller-Navarro Joint Venture Report S–N/99205–077, 643 p.

Stoller-Navarro Joint Venture, National Securities Technologies, LLC, and Lawrence Livermore National Laboratory, 2007, Phase I contaminant transport parameters for the groundwater flow and contaminant transport model of Corrective Action Unit 97: Yucca Flat/Climax Mine, Nevada Test Site, Nye County, Nevada: Stoller-Navarro Joint Venture Report S-N/99205–096, 918 p.

Stoller-Navarro Joint Venture, 2009, Unclassified source term and radionuclide data for Corrective Action Unit 97—Yucca Flat/Climax Mine, Nevada Test Site, Nevada: Stoller-Navarro Joint Venture Report S–N/99205–114, Revision No. 2, 219 p.

Stonestrom, D.A., Prudic, D.E., Laczniak, R.J., Akstin, K.C., Boyd, R.A., and Henkelman, K.K., 2003, Estimates of deep percolation beneath native vegetation, irrigated fields, and the Amargosa-River channel, Amargosa Desert, Nye County, Nevada: U.S. Geological Survey Open-File Report 03–104, 83 p.

Sweetkind, D.S., and Drake, R.M., II, 2007a, Characteristics of fault zones in volcanic rocks, Nevada Test Site, Nevada: U.S. Geological Survey Open-File Report 2007–1293, 52 p.

Sweetkind, D.S., and Drake, R.M., II, 2007b, Geologic characterization of young alluvial basin-fill deposits from borehole data in Yucca Flat, Nye County, Nevada: U.S. Geological Survey Scientific Investigations Series Report 2007–5062, 17 p.

Thordarson, William, Young, R.A., and Winograd, I.J., 1967, Records of wells and test holes in the Nevada Test Site and vicinity (through December 1966): U.S. Geological Survey Open-File Report 67-218, 26 p.

Tompson, A.F.B. (ed.), 2008, Evaluation of hydrologic source term processes for underground nuclear tests in Yucca Flat, Nevada Test Site—saturated tests: Lawrence Livermore National Laboratory Report LLNL–TR–403429 [variously paged].

Tóth, J., 1962, A theory of groundwater motion in small drainage basins in Central Alberta, Canada: Journal of Geophysical Research, v. 67, no. 11, p. 4,375–4,387.

Trexler, J.H., Jr., Cole, J.C., and Cashman, P.H., 1996, Middle Devonian through Mississippian stratigraphy on and near the Nevada Test Site—Implications for hydrocarbon potential: American Association of Petroleum Geologists Bulletin, v. 80, p. 1,736–1,762.

U.S. Congress, Office of Technology Assessment, 1989, The containment of underground nuclear explosions: U.S. Government Printing Office, Washington, DC, OTA–ISC–414, 80 p.

U.S. Department of Energy, 1997, Shaft and tunnel nuclear detonations at the Nevada Test Site: Development of a primary database for the estimation of potential interactions with the regional groundwater system, U.S. Department of Energy Report DOE/NV–464, 62 p.

U.S. Department of Energy, 2000a, Corrective action investigation plan for Corrective Action Unit 97—Yucca Flat/Climax Mine, Nevada Test Site, Nevada: U.S. Department of Energy DOE/NV–659, 381 p.

U.S. Department of Energy, 2000b, United States nuclear tests, July 1945 through September 1992: U.S. Department of Energy Report DOE/NV–209 REV 15, 162 p.

U.S. Department of Energy, 2003, Underground Test Area Project, questions and answers: U.S. Department of Energy Report DOE/NV–618, 5 p.

U.S. Geological Survey, 1983, Geologic and geophysical investigations of Climax stock intrusive, Nevada: U.S. Geological Survey Open-File Report 83-377, 82 p.

U.S. Geological Survey, 2011, USGS/U.S. Department of Energy cooperative studies in Nevada: accessed June 20, 2011 at http://nevada.usgs.gov/doe_nv/.

Waddell, R.K., Robison, J.H., and Blankennagel, R.K., 1984, Hydrology of Yucca Mountain and vicinity, Nevada–California—Investigative results through mid 1983: U.S. Geological Survey Water-Resources Investigations Report 84–4267, 72 p.

Walvoord, M.A., Phillips, F.M., Tyler, S.W., and Hartsough, P.C., 2002, Deep arid system hydrodynamics 2—Application to paleohydrologic reconstruction using vadose zone profiles from the northern Mojave Desert: Water Resources Research, v. 38, no. 12, 1291, 12 p.

Walvoord, M.A., Plummer, M.A., Phillips, F.M., and Wolfsberg, A.V., 2002, Deep arid system hydrodynamics 1—Equilibrium states and response times in thick desert vadose zones: Water Resources Research, v. 38, no. 12, 1308, 15 p.

Western Regional Climate Center, 2012, Standardized precipitation index: accessed May 2012, at http://www.wrcc.dri.edu/spi.

Winograd, I.J., and Pearson F.J., Jr., 1976, Major carbon 14 anomaly in a regional carbonate aquifer—Possible evidence for megascale channeling, south-central Great Basin: Water Resources Research, v. 12, p. 1,125–1,143.

Winograd, I.J., Riggs, A.C., and Coplen, T.B., 1998, The relative contributions of summer and cool-season precipitation to groundwater recharge, Spring Mountains, Nevada, USA: Hydrogeology Journal, v. 6, p. 77–93.

Winograd, I.J., and Thordarson, William, 1975, Hydrogeologic and hydrochemical framework, south-central Great Basin, Nevada–California, with special reference to the Nevada Test Site: U.S. Geological Survey Professional Paper 712–C, 126 p.

Wohletz, Kenneth, Wolfsberg, Andrew, Olson, Alyssa, and Gable, Carl, 1999, Evaluating the effects of underground nuclear testing below the water table on groundwater and radionuclide migration in the Tuff Pile I region of Yucca Flat—Numerical simulations: Los Alamos National Laboratory, EES–1 UGTA FY99 Report, 31 p.

Wood, D.B., 2007, Digitally available interval-specific rock-sample data compiled from historical records, Nevada Test Site and vicinity, Nye County, Nevada: U.S. Geological Survey Data Series 297, version 2.0, October 2009, 58 p.

Appendix 1.

Appendix 1.   Water Levels Measured in Yucca Flat, Nevada National Security Site, 1951–2010

Hydrographs and locations for the 229 wells that have measured water levels in Yucca Flat are tabulated and can be displayed interactively from a Microsoft® Excel workbook. The workbook is designed to be an easy-to-use tool to view water levels and other associated information for wells in the study area. Information for an individual well can be selected by using the AutoFilter option available in Excel. To select a well, click the filter arrow beside the "Well name" column heading and Excel will display a dropdown list of all wells in the spreadsheet. Deselect (uncheck) the currently selected well and select (check) the well of interest. Water-level information for the new well will be displayed. The information presented for a selected well includes the following:

- USGS site identification number
- Well name
- Land-surface altitude
- Water-level date
- Water-level depth
- Water-level altitude
- Water-level qualifier
- Water-level source
- Water-level status
- Water-level method
- Water-level accuracy
- Water-level remark
- Predevelopment use flag
- Transient nuclear test flag
- Transient pumping flag
- Latitude
- Longitude

Appendix 1 data are available at *http://pubs.usgs.gov/sir/2012/5196/*.

Appendix 2.

Appendix 2. Well Characteristics, Hydraulic Heads, and Identification of Wells with Water Levels Representative of Predevelopment Conditions or Affected by Nuclear Testing or Pumping in Yucca Flat, Nevada National Security Site.

A summary table that includes the 229 wells with measured water levels in Yucca Flat is available in a Microsoft® Excel workbook. For each well, the mean of the water levels considered representative of predevelopment conditions and the calculated hydraulic head are presented. The information presented for each well includes:

- Well name
- USGS site identification number
- Borehole name, as used in report
- Nevada National Security Site Red Book hole number
- NNSS area number
- Latitude
- Longitude
- Land-surface altitude
- Land-surface altitude accuracy
- Depth drilled
- Well depth
- Top and bottom opening altitude
- Number of water levels
- Water-level date range
- Hydraulic-head estimate
- Water-column length
- Contributing subsurface hydrologic unit types
- Thicknesses of contributing subsurface hydrologic unit types
- Contributing hydrostratigraphic units
- Thicknesses of contributing hydrostratigraphic units
- Predevelopment map use of hydraulic head
- Predevelopment water-level certainty
- Transient nuclear test water-level certainty
- Transient pumping water-level certainty

Appendix 2 data are available at *http://pubs.usgs.gov/sir/2012/5196/.*

Appendix 3.

Appendix 3.   Hydrostratigraphic Units and Subsurface Hydrologic Unit Types for Wells and Underground Nuclear Test Holes in Yucca Flat, as Projected From Hydrostratigraphic Framework Models.

The hydrostratigraphic units (HSUs) and corresponding subsurface hydrologic unit types (SHUTs) are tabulated for 941 sites in Yucca Flat that consist of (1) wells having measured water levels and (2) boreholes where underground nuclear tests were conducted. These sites can be displayed interactively from a Microsoft® Excel macro-driven workbook. The workbook is designed to view a stratigraphic column interpreted from a hydrostratigraphic framework model, the mean predevelopment water-level altitude, and basic well-construction information for wells in the study area. Information for an individual well or borehole can be viewed by selecting it from a column-header dropdown list.

Appendix 3 data are available at
http://pubs.usgs.gov/sir/2012/5196/.

USGS

Fenelon and others—**Conceptualization of the Predevelopment Groundwater Flow System and Transient Water-Level Responses in Yucca Flat, Nevada National Security Site, Nevada**—Scientific Investigations Report 2012-5196

www.ingramcontent.com/pod-product-compliance
Lightning Source LLC
Chambersburg PA
CBHW081848170526
45167CB00007B/2932